LOVE'S MEINIE: THREE LECTURES ON GREEK AND ENGLISH BIRDS

JOHN RUSKIN

ZINC READ

Published by
This edition copyright © 2024 by Zinc Read
Design and setting By Zinc Read
Email- zincread@gmail.com
ISBN-978-93-6277-618-1

CONTENTS

PREFACE.

BRANTWOOD, 9*th June*, 1881.
Quarter past five, morning.

The birds chirping feebly,--mostly chaffinches answering each other, the rest discomposed, I fancy, by the June snow;[1] the lake neither smooth nor rippled, but like a surface of perfectly bright glass, ill cast; the lines of wave few and irregular, like flaws in the planes of a fine crystal.

I see this book was begun eight years ago;--then intended to contain only four Oxford lectures: but the said lectures also 'intended' to contain the cream of forty volumes of scientific ornithology. Which intentions, all and sundry, having gone, Carlyle would have said, to water, and more piously-minded persons, to fire, I am obliged now to cast my materials into another form: and here, at all events, is a bundle of what is readiest under my hand. The nature and name of which I must try to make a little more intelligible than my books have lately been, either in text or title.

'Meinie' is the old English word for 'Many,' in the sense of 'a many' persons attending one, as bridesmaids, when in sixes or tens or dozens;--courtiers, footmen, and the like. It passes gradually into 'Menial,' and unites the senses of Multitude and Servitude.

In the passages quoted from, or referred to in, Chaucer's translation of the Romance of the Rose, at the end of the first lecture, any reader who cares for a clue to the farther significances of the title, may find one to lead him safely through richer labyrinths of thought than mine: and ladder enough also,--if there be either any heavenly, or pure earthly, Love, in his own breast,--to guide him to a pretty bird's nest; both in the Romances of the Rose and of Juliet, and in the Sermons of St. Francis and St. Bernard.

The term 'Lecture' is retained, for though I lecture no more, I still write habitually in a manner suited for oral delivery, and imagine myself speaking to my pupils, if ever I am happily thinking in myself. But it will be also seen that by the help of this very familiarity of style, I am endeavoring, in these and my other writings on Natural History, to compel in the student a clearness of thought and precision of language which have not hitherto been in any wise the virtues, or skills, of scientific persons. Thoughtless readers, who imagine that my own style (such as it is, the one thing which the British public concedes to me as a real power) has been formed without pains, may smile at the confidence with which I speak of altering accepted, and even long-established, nomenclature. But the use which I now have of language has taken me forty years to attain; and those forty years spent, mostly, in walking through the wilderness of this world's vain words, seeking how they might be pruned into some better strength. And I think it likely that at last I may put in my pruning-hook with effect; for indeed a time must come when English fathers and mothers will wish their children to learn English again, and to speak it for all scholarly purposes; and, if they use, instead, Greek or Latin, to use them only that they may be understood by Greeks or Latins;[2] and not that they may mystify the illiterate many of their own land. Dead languages, so called, may at least be left at rest, if not honored; and must not be torn in mutilation out of their tumuli, that the skins and bones of them may help to hold our living nonsense together; while languages called living, but which live only to slack themselves into slang, or bloat themselves into bombast, must one day have new grammars written for their license, and new laws for their insolence.

Observe, however, that the recast methods of classification adopted in this book, and in 'Proserpina,' must be carefully distinguished from their recastings of nomenclature. I am perfectly sure that it is wiser to use plain short words than obscure long ones; but not in the least sure that I am doing the best that can be done for my pupils, in classing swallows with owls, or milkworts with violets. The classification is always given as tentative; and, at its utmost, elementary: but the nomenclature, as in all probability conclusive.

For the rest, the success and the service of all depend on the more or less thorough accomplishment of plans long since laid, and which would have been good for little if their coping could at once have been conjectured or foretold in their foundations. It has been throughout my trust, that if Death should write on these, "What this man began to build, he was not able to finish," God may also write on them, not in anger, but in aid,

"A stronger than he, cometh."

LOVE'S MEINIE.

"Il etoit tout convert d'oisiaulx."

Romance of the Rose.

LECTURE I.

THE ROBIN.

1. Among the more splendid pictures in the Exhibition of the Old Masters, this year, you cannot but remember the Vandyke portraits of the two sons of the Duke of Lennox. I think you cannot but remember it, because it would be difficult to find, even among the works of Vandyke, a more striking representation of the youth of our English noblesse; nor one in which the painter had more exerted himself, or with better success, in rendering the decorous pride and natural grace of honorable aristocracy.

Vandyke is, however, inferior to Titian and Velasquez, in that his effort to show this noblesse of air and persons may always be detected; also the aristocracy of Vandyke's day were already so far fearful of their own position as to feel anxiety that it should be immediately recognized. And the effect of the painter's conscious deference, and of the equally conscious pride of the boys, as they stood to be painted, has been somewhat to shorten the power of the one, and to abase the dignity of the other. And thus, in the midst of my admiration of the youths' beautiful faces, and natural quality of majesty, set off by all splendors of dress and courtesies of art, I could not forbear questioning with myself what the true value was, in the scales of creation, of these fair human beings who set so high a value on themselves; and,--as if the only answer,--the words kept repeating themselves in my ear, "Ye are of more value than many sparrows."

2. Passeres, stronthos [Greek: strouthos]--the things that open their wings, and are not otherwise noticeable; small birds of the land and wood; the food of the serpent, of man, or of the stronger creatures of their own kind,--that even these, though among the simplest and obscurest of beings, have yet price in the eyes of their Maker, and that the death of one of them cannot take place but by His permission, has long been the subject of declamation in our pulpits, and the ground of much sentiment in nursery education. But the declamation is so aimless, and the sentiment so hollow, that, practically, the chief interest of the leisure of mankind has been found in the destruction of the creatures which they professed to believe even the Most High would not see perish without pity; and, in recent days, it is fast becoming the only definition of aristocracy, that the principal business of its life is the killing of sparrows.

Sparrows, or pigeons, or partridges, what does it matter? "Centum mille perdrices plumbo confecit;"[4] that is, indeed, too often the sum of the life of an English lord; much questionable now, if *indeed* of more value than that of many sparrows.

3. Is it not a strange fact, that, interested in nothing so much for the last two hundred years, as in his horses, he yet left it to the farmers of Scotland to relieve draught horses from the bearing-rein?[5] Is it not one equally strange that, master of the forests of England for a thousand years, and of its libraries for three hundred, he left the natural history of birds to be written by a card-printer's lad of Newcastle?[6] Written, and not written, for indeed we have no natural history of birds written yet. It cannot be written but by a scholar and a gentleman; and no English gentleman in recent times has ever thought of birds except as flying targets, or flavorous dishes. The only piece of natural history worth the name in the English language, that I know of, is in the few lines of Milton on the Creation. The only example of a proper manner of contribution to natural history is in White's Letters from Selborne. You know I have always spoken of Bewick as pre-eminently a vulgar or boorish person, though of splendid honor and genius; his vulgarity shows in nothing so much as in the poverty of the details he has collected, with the best intentions, and the shrewdest sense, for English ornithology. His imagination is not cultivated enough to enable him to choose, or arrange.

4. Nor can much more be said for the observations of modern science. It is vulgar in a far worse way, by its arrogance and materialism. In general, the scientific natural history of a bird consists of four articles,--first, the name and estate of the gentleman whose gamekeeper shot the

last that was seen in England; secondly, two or three stories of doubtful origin, printed in every book on the subject of birds for the last fifty years; thirdly, an account of the feathers, from the comb to the rump, with enumeration of the colors which are never more to be seen on the living bird by English eyes; and, lastly, a discussion of the reasons why none of the twelve names which former naturalists have given to the bird are of any further use, and why the present author has given it a thirteenth, which is to be universally, and to the end of time, accepted.

5. You may fancy this is caricature; but the abyss of confusion produced by modern science in nomenclature, and the utter void of the abyss when you plunge into it after any one useful fact, surpass all caricature. I have in my hand thirteen plates of thirteen species of eagles; eagles all, or hawks all, or falcons all--whichever name you choose for the great race of the hook-headed birds of prey--some so like that you can't tell the one from the other, at the distance at which I show them to you, all absolutely alike in their eagle or falcon character, having, every one, the falx for its beak, and every one, flesh for its prey. Do you suppose the unhappy student is to be allowed to call them all eagles, or all falcons, to begin with, as would be the first condition of a wise nomenclature, establishing resemblance by specific name, before marking variation by individual name? No such luck. I hold you up the plates of the thirteen birds one by one, and read you their names off the back:--

The first, is
an Aquila.
The second,
a Haliaetus.
The third,
a Milvus.
The fourth,
a Pandion.
The fifth,
an Astur.
The sixth,
a Falco.
The seventh,
a Pernis.
The eighth,
a Circus.
The ninth,

a Buteo.
The tenth,
an Archibuteo.
The eleventh,
an Accipiter.
The twelfth,
an Erythropus.
And the thirteenth,
a Tinnunculus.

There's a nice little lesson to entertain a parish school-boy with, beginning his natural history of birds!

6. There are not so many varieties of robin as of hawk, but the scientific classifiers are not to be beaten. If they cannot find a number of similar birds to give different names to, they will give two names to the same one. Here are two pictures of your own redbreast, out of the two best modern works on ornithology. In one, it is called "Motacilla rubecula;" in the other, "Rubecula familiaris."

7. It is indeed one of the most serious, as one of the most absurd, weaknesses, of modern naturalists to imagine that *any* presently invented nomenclature can stand, even were it adopted by the consent of nations, instead of the conceit of individuals. It will take fifty years' digestion before the recently ascertained elements of natural science can permit the arrangement of species in any permanently (even over a limited period) namable order; nor then, unless a great man is born to perceive and exhibit such order. In the meantime, the simplest and most descriptive nomenclature is the best. Every one of these birds, for instance, might be called falco in Latin, hawk in English, some word being added to distinguish the genus, which should describe its principal aspect or habit. Falco montium, Mountain Hawk; Falco silvarum, Wood Hawk; Falco procellarum, Sea Hawk; and the like. Then, one descriptive epithet would mark species. Falco montium, aureus, Golden Eagle; Falco silvarum, apivorus, Honey Buzzard; and so on; and the naturalists of Vienna, Paris, and London should confirm the names of known creatures, in conclave, once every half-century, and let them so stand for the next fifty years.

8. In the meantime, you yourselves, or, to speak more generally, the young rising scholars of England,--all of you who care for life as well as literature, and for spirit,--even the poor souls of birds,--as well as lettering of their classes in books,--you, with all care, should cherish the old Saxon-English and Norman-French names of birds, and ascertain

them with the most affectionate research--never despising even the rudest or most provincial forms: all of them will, some day or other, give you clue to historical points of interest. Take, for example, the common English name of this low-flying falcon, the most tamable and affectionate of his tribe, and therefore, I suppose, fastest vanishing from field and wood, the buzzard. That name comes from the Latin "buteo," still retained by the ornithologists; but, in its original form, valueless, to you. But when you get it comfortably corrupted into Provencal "Busac," (whence gradually the French busard, and our buzzard,) you get from it the delightful compound "busacador," "adorer of buzzards"--meaning, generally, a sporting person; and then you have Dante's Bertrand de Born, the first troubadour of war, bearing witness to you how the love of mere hunting and falconry was already, in his day, degrading the military classes, and, so far from being a necessary adjunct of the noble disposition of lover or soldier, was, even to contempt, showing itself separate from both.

"Le ric home, cassador,
M'enneion, e'l buzacador.
Parlan de volada, d'austor,
Ne jamais, d'armas, ni d'amor."
The rich man, the chaser,
Tires me to death; and the adorer of buzzards.
They talk of covey and hawk,
And never of arms, nor of love.

"Cassador," of course, afterwards becomes "chasseur," and "austor" "vautour." But after you have read this, and familiarized your ear with the old word, how differently Milton's phrase will ring to you,--"Those who thought no better of the Living God than of a buzzard idol,"--and how literal it becomes, when we think of the actual difference between a member of Parliament in Milton's time, and the Busacador of to-day;--and all this freshness and value in the reading, observe, come of your keeping the word which great men have used for the bird, instead of letting the anatomists blunder out a new one from their Latin dictionaries.

9. There are not so many namable varieties, I just now said, of robin as of falcon; but this is somewhat inaccurately stated. Those thirteen birds represented a very large proportion of the entire group of the birds of prey, which in my sevenfold classification I recommended you to call universally, "hawks." The robin is only one of the far greater multitude of small birds which live almost indiscriminately on grain or insects, and

which I recommended you to call generally "sparrows"; but of the robin itself, there are two important European varieties--one redbreasted, and the other blue-breasted.

10. You probably, some of you, never heard of the blue-breast; very few, certainly, have seen one alive, and, if alive, certainly not wild in England.

Here is a picture of it, daintily done,[7] and you can see the pretty blue shield on its breast, perhaps, at this distance. Vain shield, if ever the fair little thing is wretched enough to set foot on English ground! I find the last that was seen was shot at Margate so long ago as 1842,--and there seems to be no official record of any visit before that, since Mr. Thomas Embledon shot one on Newcastle town moor in 1816. But this rarity of visit to us is strange; other birds have no such clear objection to being shot, and really seem to come to England expressly for the purpose. And yet this blue-bird--(one can't say "blue robin"--I think we shall have to call him "bluet," like the cornflower)--stays in Sweden, where it sings so sweetly that it is called "a hundred tongues."

11. That, then, is the utmost which the lords of land, and masters of science, do for us in their watch upon our feathered suppliants. One kills them, the other writes classifying epitaphs.

We have next to ask what the poets, painters, and monks have done.

The poets--among whom I affectionately and reverently class the sweet singers of the nursery, mothers and nurses--have done much; very nearly all that I care for your thinking of. The painters and monks, the one being so greatly under the influence of the other, we may for the present class together; and may almost sum their contributions to ornithology in saying that they have plucked the wings from birds, to make angels of men, and the claws from birds, to make devils of men.

If you were to take away from religious art these two great helps of its--I must say, on the whole, very feeble--imagination; if you were to take from it, I say, the power of putting wings on shoulders, and claws on fingers and toes, how wonderfully the sphere of its angelic and diabolic characters would be contracted! Reduced only to the sources of expression in face or movements, you might still find in good early sculpture very sufficient devils; but the best angels would resolve themselves, I think, into little more than, and not often into so much as, the likenesses of pretty women, with that grave and (I do not say it ironically) majestic expression which they put on, when, being very fond of their husbands and children, they seriously think either the one or the other have misbehaved themselves.

12. And it is not a little discouraging for me, and may well make you doubtful of my right judgment in this endeavor to lead you into closer attention to the bird, with its wings and claws still in its own possession;--it is discouraging, I say, to observe that the beginning of such more faithful and accurate observation in former art, is exactly coeval with the commencement of its decline. The feverish and ungraceful natural history of Paul, called, "of the birds," Paolo degli Uccelli, produced, indeed, no harmful result on the minds of his contemporaries, they watched in him, with only contemptuous admiration, the fantasy of zoological instinct which filled his house with painted dogs, cats, and birds, because he was too poor to fill it with real ones. Their judgment of this morbidly naturalistic art was conclusively expressed by the sentence of Donatello, when going one morning into the Old Market, to buy fruit, and finding the animal painter uncovering a picture, which had cost him months of care, (curiously symbolic in its subject, the infidelity of St. Thomas, of the investigatory fingering of the natural historian,) "Paul, my friend," said Donatello, "thou art uncovering the picture just when thou shouldst be shutting it up."

13. No harm, therefore, I repeat, but, on the contrary, some wholesome stimulus to the fancy of men like Luca and Donatello themselves, came of the grotesque and impertinent zoology of Uccello.

But the fatalest institutor of proud modern anatomical and scientific art, and of all that has polluted the dignity, and darkened the charity, of the greater ages, was Antonio Pollajuolo of Florence. Antonio (that is to say) the Poulterer--so named from the trade of his grandfather, and with just so much of his grandfather's trade left in his own disposition, that being set by Lorenzo Ghiberti to complete one of the ornamental festoons of the gates of the Florentine Baptistery, there, (says Vasari) "Antonio produced a quail, which may still be seen, and is so beautiful, nay, so perfect, that it wants nothing but the power of flight."

14. Here, the morbid tendency was as attractive as it was subtle. Ghiberti himself fell under the influence of it; allowed the borders of his gates, with their fluttering birds and bossy fruits, to dispute the spectators' favor with the religious subjects they inclosed; and, from that day forward, minuteness and muscularity were, with curious harmony of evil, delighted in together; and the lancet and the microscope, in the hands of fools, were supposed to be complete substitutes for imagination in the souls of wise men: so that even the best artists are gradually compelled, or beguiled, into compliance with the curiosity of their day; and Francia, in the city of Bologna, is held to be a "kind of god, more

particularly" (again I quote Vasari) "after he had painted a set of caparisons for the Duke of Urbino, on which he depicted a great forest all on fire, and whence there rushes forth an immense number of every kind of animal, with several human figures. This terrific, yet truly beautiful representation, was all the more highly esteemed for the time that had been expended on it in the plumage of the birds, and other minutiae in the delineation of the different animals, and in the diversity of the branches and leaves of the various trees seen therein;" and thenceforward the catastrophe is direct, to the ornithological museums which Breughel painted for gardens of Eden, and to the still life and dead game of Dutch celebrities.

15. And yet I am going to invite you to-day to examine, down to almost microscopic detail, the aspect of a small bird, and to invite you to do this, as a most expedient and sure step in your study of the greatest art.

But the difference in our motive of examination will entirely alter the result. To paint birds that we may show how minutely we can paint, is among the most contemptible occupations of art. To paint them, that we may show how beautiful they are, is not indeed one of its highest, but quite one of its pleasantest and most useful; it is a skill within the reach of every student of average capacity, and which, so far as acquired, will assuredly both make their hearts kinder, and their lives happier.

Without further preamble, I will ask you to look to-day, more carefully than usual, at your well-known favorite, and to think about him with some precision.

16. And first, Where does he come from? I stated that my lectures were to be on English and Greek birds; but we are apt to fancy the robin all our own. How exclusively, do you suppose, he really belongs to us? You would think this was the first point to be settled in any book about him. I have hunted all my books through, and can't tell you how much he is our own, or how far he is a traveler.

And, indeed, are not all our ideas obscure about migration itself? You are broadly told that a bird travels, and how wonderful it is that it finds its way; but you are scarcely ever told, or led to think, what it really travels for--whether for food, for warmth, or for seclusion--and how the traveling is connected with its fixed home. Birds have not their town and country houses,--their villas in Italy, and shooting boxes in Scotland. The country in which they build their nests is their proper home,--the country, that is to say, in which they pass the spring and summer. Then they go south in the winter, for food and warmth; but in what lines, and by what stages? The general definition of a migrant in this hemisphere is a bird that goes north to build its nest, and south for the winter; but, then, the one essential point to know about it is the breadth and latitude of the zone it properly inhabits,--that is to say, in which it builds its nest; next, its habits of life, and extent and line of southing in the winter; and finally, its manner of traveling.

17. Now, here is this entirely familiar bird, the robin. Quite the first thing that strikes me about it, looking at it as a painter, is the small effect it seems to have had on the minds of the southern nations. I trace nothing of it definitely, either in the art or literature of Greece or Italy. I find, even, no definite name for it; you don't know if Lesbia's "passer" had a red breast, or a blue, or a brown. And yet Mr. Gould says it is abundant in all parts of Europe, in all the islands of the Mediterranean, and in Madeira and the Azores. And then he says--(now notice the puzzle of this),--"In many parts of the Continent it is a migrant, and, contrary to what obtains with us, is there treated as a vagrant, for there is scarcely a country across the water in which it is not shot down and eaten."

"In many parts of the Continent it is a migrant." In what parts--how far--in what manner?

18. In none of the old natural history books can I find any account of the robin as a traveler, but there is, for once, some sufficient reason for their reticence. He has a curious fancy in his manner of traveling. Of all birds, you would think he was likely to do it in the cheerfulest way, and he does it in the saddest. Do you chance to have read, in the Life of Charles Dickens, how fond he was of taking long walks in the night and alone? The robin, en voyage, is the Charles Dickens of birds. He always travels in the night, and alone; rests, in the day, wherever day chances to find him; sings a little, and pretends he hasn't been anywhere. He goes as far, in the winter, as the north-west of Africa; and in Lombardy, arrives from the south early in March; but does not stay long, going on into the Alps, where he prefers wooded and wild districts. So, at least, says my Lombard informant.

I do not find him named in the list of Cretan birds; but even if often seen, his dim red breast was little likely to make much impression on the Greeks, who knew the flamingo, and had made it, under the name of Phoenix or Phoenicopterus, the center of their myths of scarlet birds. They broadly embraced the general aspect of the smaller and more obscure species, under the term xonthos [Greek: xonthos], which, as I understand their use of it, exactly implies the indescribable silky brown, the groundwork of all other color in so many small birds, which is indistinct among green leaves, and absolutely identifies itself with dead ones, or with mossy stems.

19. I think I show it you more accurately in the robin's back than I could in any other bird; its mode of transition into more brilliant color is, in him, elementarily simple; and although there is nothing, or rather because there is nothing, in his plumage, of interest like that of tropical birds, or even of our own game-birds, I think it will be desirable for you to learn first from the breast of the robin what a feather is. Once knowing that, thoroughly, we can further learn from the swallow what a wing is; from the chough what a beak is; and from the falcon what a claw is.

I must take care, however, in neither of these last two particulars, to do injustice to our little English friend here; and before we come to his feathers, must ask you to look at his bill and his feet.

20. I do not think it is distinctly enough felt by us that the beak of a bird is not only its mouth, but its hand, or rather its two hands. For, as its arms and hands are turned into wings, all it has to depend upon, in economical and practical life, is its beak. The beak, therefore, is at once its sword, its carpenter's tool-box, and its dressing-case; partly also its musical instrument; all this besides its function of seizing and preparing the food, in which functions alone it has to be a trap, carving-knife, and teeth, all in one.

21. It is this need of the beak's being a mechanical tool which chiefly regulates the form of a bird's face, as opposed to a four-footed animal's. If the question of food were the only one, we might wonder why there were not more four-footed creatures living on seeds than there are; or why those that do--field-mice and the like--have not beaks instead of teeth. But the fact is that a bird's beak is by no means a perfect eating or food-seizing instrument. A squirrel is far more dexterous with a nut than a cockatoo; and a dog manages a bone incomparably better than an eagle. But the beak has to do so much more! Pruning feathers, building nests, and the incessant discipline in military arts, are all to be thought of, as much as feeding.

Soldiership, especially, is a much more imperious necessity among birds than quadrupeds. Neither lions nor wolves habitually use claws or teeth in contest with their own species; but birds, for their partners, their nests, their hunting-grounds, and their personal dignity, are nearly always in contention; their courage is unequaled by that of any other race of animals capable of comprehending danger; and their pertinacity and endurance have, in all ages, made them an example to the brave, and an amusement to the base, among mankind.

22. Nevertheless, since as sword, as trowel, or as pocket-comb, the beak of the bird has to be pointed, the collection of seeds may be conveniently intrusted to this otherwise penetrative instrument, and such food as can only be obtained by probing crevices, splitting open fissures, or neatly and minutely picking things up, is allotted, pre-eminently, to the bird species.

The food of the robin, as you know, is very miscellaneous. Linnaeus says of the Swedish one, that it is "delectatus euonymi baccis,"--"delighted with dogwood berries,"--the dogwood growing abundantly in Sweden, as once in Forfarshire, where it grew, though only a bush usually in the south, with trunks a foot or eighteen inches in diameter, and the tree thirty feet high. But the Swedish robin's taste for its berries is to be noted by you, because, first, the dogwood berry is commonly said to be so bitter that it is not eaten by birds (Loudon, "Arboretum," ii., 497, 1.); and, secondly, because it is a pretty coincidence that this most familiar of household birds should feed fondly from the tree which gives the housewife her spindle,--the proper name of the dogwood in English, French, and German being alike "Spindle-tree." It feeds, however, with us, certainly, most on worms and insects. I am not sure how far the following account of its mode of dressing its dinners may be depended on: I take it from an old book on Natural History, but find it, more or less, confirmed by others: "It takes a worm by one extremity in its beak, and beats it on the ground till the inner part comes away. Then seizing it in a similar manner by the other end, it entirely cleanses the outer part, which alone it eats."

One's first impression is that this must be a singularly unpleasant operation for the worm, however fastidiously delicate and exemplary in the robin. But I suppose the real meaning is, that as a worm lives by passing earth through its body, the robin merely compels it to quit this--not ill-gotten, indeed, but now quite unnecessary--wealth. We human creatures, who have lived the lives of worms, collecting dust, are served by Death in exactly the same manner.

23. You will find that the robin's beak, then, is a very prettily representative one of general bird power. As a weapon, it is very formidable indeed; he can kill an adversary of his own kind with one blow of it in the throat; and is so pugnacious, "valde pugnax," says Linnaeus, "ut non una arbor duos capiat erithacos,"--"no single tree can hold two cock-robins;" and for precision of seizure, the little flat hook at

the end of the upper mandible is one of the most delicately formed points of forceps which you can find among the grain eaters. But I pass to one of his more special perfections.

24. He is very notable in the exquisite silence and precision of his movements, as opposed to birds who either creak in flying, or waddle in walking. "Always quiet," says Gould, "for the silkiness of his plumage renders his movements noiseless, and the rustling of his wings is never heard, any more than his tread on earth, over which he bounds with amazing sprightliness." You know how much importance I have always given, among the fine arts, to good dancing. If you think of it, you will find one of the robin's very chief ingratiatory faculties is his dainty and delicate movement,--his footing it featly here and there. Whatever prettiness there may be in his red breast, at his brightest he can always be outshone by a brickbat. But if he is rationally proud of anything about him, I should think a robin must be proud of his legs. Hundreds of birds have longer and more imposing ones--but for real neatness, finish, and precision of action, commend me to his fine little ankles, and fine little feet; this long stilted process, as you know, corresponding to our ankle-bone. Commend me, I say, to the robin for use of his ankles--he is, of all birds, the pre-eminent and characteristic Hopper; none other so light, so pert, or so swift.

25. We must not, however, give too much credit to his legs in this matter. A robin's hop is half a flight; he hops, very essentially, with wings and tail, as well as with his feet, and the exquisitely rapid opening and quivering of the tail-feathers certainly give half the force to his leap. It is in this action that he is put among the motacillae, or wagtails; but the ornithologists have no real business to put him among them. The swing of the long tail feathers in the true wagtail is entirely consequent on its motion, not impulsive of it--the tremulous shake is *after* alighting. But the robin leaps with wing, tail, and foot, all in time, and all helping each other. Leaps, I say; and you check at the word; and ought to check: you look at a bird hopping, and the motion is so much a matter of course, you never think how it is done. But do you think you would find it easy to hop like a robin if you had two--all but wooden--legs, like this?

26. I have looked wholly in vain through all my books on birds, to find some account of the muscles it uses in hopping, and of the part of the toes with which the spring is given. I must leave you to find out that for yourselves; it is a little bit of anatomy which I think it highly desirable for you to know, but which it is not my business to teach you. Only observe, this is the point to be made out. You leap yourselves, with the

toe and ball of the foot; but, in that power of leaping, you lose the faculty of grasp; on the contrary, with your hands, you grasp as a bird with its feet. But you cannot hop on your hands. A cat, a leopard, and a monkey, leap or grasp with equal ease; but the action of their paws in leaping is, I imagine, from the fleshy ball of the foot; while in the bird, characteristically gamphonux [Greek: gampsonux], this fleshy ball is reduced to a boss or series of bosses, and the nails are elongated into sickles or horns; nor does the springing power seem to depend on the development of the bosses. They are far more developed in an eagle than a robin; but you know how unpardonably and preposterously awkward an eagle is when he hops. When they are most of all developed, the bird walks, runs, and digs well, but leaps badly.

27. I have no time to speak of the various forms of the ankle itself, or of the scales of armor, more apparent than real, by which the foot and ankle are protected. The use of this lecture is not either to describe or to exhibit these varieties to you, but so to awaken your attention to the real points of character, that, when you have a bird's foot to draw, you may do so with intelligence and pleasure, knowing whether you want to express force, grasp, or firm ground pressure, or dexterity and tact in motion. And as the actions of the foot and the hand in man are made by every great painter perfectly expressive of the character of mind, so the expressions of rapacity, cruelty, or force of seizure, in the harpy, the gryphon, and the hooked and clawed evil spirits of early religious art, can only be felt by extreme attention to the original form.

28. And now I return to our main question, for the robin's breast to answer, "What is a feather?" You know something about it already; that it is composed of a quill, with its lateral filaments terminating generally, more or less, in a point; that these extremities of the quills, lying over each other like the tiles of a house, allow the wind and rain to pass over them with the least possible resistance, and form a protection alike from the heat and the cold; which, in structure much resembling the scale-armor assumed by man for very different objects, is, in fact, intermediate, exactly, between the fur of beasts and the scales of fishes; having the minute division of the one, and the armor-like symmetry and succession of the other.

29. Not merely symmetry, observe, but extreme flatness. Feathers are smoothed down, as a field of corn by wind with rain; only the swathes laid in beautiful order. They are fur, so structurally placed as to imply, and submit to, the perpetually swift forward motion. In fact, I have no doubt the Darwinian theory on the subject is that the feathers of birds

once stuck up all erect, like the bristles of a brush, and have only been blown flat by continual flying.

Nay, we might even sufficiently represent the general manner of conclusion in the Darwinian system by the statement that if you fasten a hair-brush to a mill-wheel, with the handle forward, so as to develop itself into a neck by moving always in the same direction, and within continual hearing of a steam-whistle, after a certain number of revolutions the hair-brush will fall in love with the whistle; they will marry, lay an egg, and the produce will be a nightingale.

30. Whether, however, a hog's bristle can turn into a feather or not, it is vital that you should know the present difference between them.

The scientific people will tell you that a feather is composed of three parts--the down, the laminae, and the shaft.

But the common-sense method of stating the matter is that a feather is composed of two parts, a shaft with lateral filaments. For the greater part of the shaft's length, these filaments are strong and nearly straight, forming, by their attachment, a finely warped sail, like that of a windmill. But towards the root of the feather they suddenly become weak, and confusedly flexible, and form the close down which immediately protects the bird's body.

To show you the typical arrangement of these parts, I choose, as I have said, the robin; because, both in his power of flying, and in his color, he is a moderate and balanced bird;--not turned into nothing but wings, like a swallow, or nothing but neck and tail, like a peacock. And first for his flying power. There is one of the long feathers of robin's wing, and here (Fig. 1) the analysis of its form.

31. First, in pure outline (A), seen from above, it is very nearly a long oval, but with this peculiarity, that it has, as it were, projecting shoulders at *a* 1 and *a* 2. I merely desire you to observe this, in passing, because one usually thinks of the contour as sweeping unbroken from the root to the point. I have not time to-day to enter on any discussion of the reason for it, which will appear when we examine the placing of the wing feathers for their stroke.

Now, I hope you are getting accustomed to the general method in which I give you the analysis of all forms--leaf, or feather, or shell, or limb. First, the plan; then the profile; then the cross-section.

I take next, the profile of my feather (B, Fig. 1), and find that it is twisted as the sail of a windmill is, but more distinctly, so that you can always see the upper surface of the feather at its root, and the under at its end. Every primary wing-feather, in the fine flyers, is thus twisted; and is best described as a sail striking with the power of a cimeter, but with the flat instead of the edge.

FIG. 1.

(Twice the size of reality.)

32. Further, you remember that on the edges of the broad side of feathers you find always a series of undulations, irregularly sequent, and lapping over each other like waves on sand. You might at first imagine that this appearance was owing to a slight ruffling or disorder of the filaments; but it is entirely normal, and, I doubt not, so constructed, in order to insure a redundance of material in the plume, so that no accident or pressure from wind may leave a gap anywhere. How this redundance is obtained you will see in a moment by bending any feather the wrong way. Bend, for instance, this plume, B, Fig. 2, into the reversed curve, A, Fig. 2; then all the filaments of the plume become perfectly even, and there are no waves at the edge. But let the plume return into its proper form, B, and the tissue being now contracted into a smaller space, the edge waves are formed in it instantly.

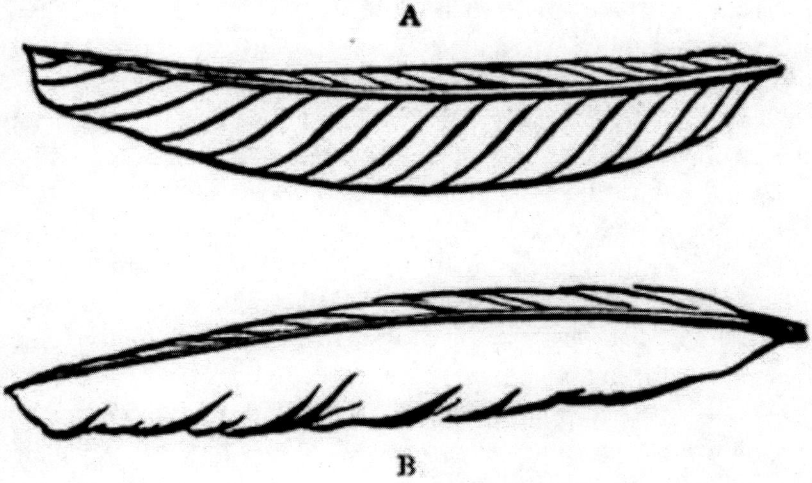

FIG. 2.

Hitherto, I have been speaking only of the filaments arranged for the strength and continuity of the energetic plume; they are entirely different when they are set together for decoration instead of force. After the feather of the robin's wing, let us examine one from his breast.

33. I said, just now, he might be at once outshone by a brickbat. Indeed, the day before yesterday, sleeping at Lichfield, and seeing, the first thing when I woke in the morning, (for I never put down the blinds of my bedroom windows,) the not uncommon sight in an English country town of an entire house-front of very neat, and very flat, and very red bricks, with very exactly squared square windows in it; and not feeling myself in anywise gratified or improved by the spectacle, I was thinking how in this, as in all other good, the too much destroyed all. The breadth of a robin's breast in brick-red is delicious, but a whole house-front of brick-red as vivid, is alarming. And yet one cannot generalize even that trite moral with any safety--for infinite breadth of green is delightful, however green; and of sea or sky, however blue.

You must note, however, that the robin's charm is greatly helped by the pretty space of gray plumage which separates the red from the brown back, and sets it off to its best advantage. There is no great brilliancy in it, even so relieved; only the finish of it is exquisite.

34. If you separate a single feather, you will find it more like a transparent hollow shell than a feather (so delicately rounded the surface of it),--gray at the root, where the down is,--tinged, and only tinged, with red at the part that overlaps and is visible; so that, when three or four more feathers have overlapped it again, all together, with their joined red, are just enough to give the color determined upon, each of them contributing a tinge. There are about thirty of these glowing filaments on each side, (the whole being no larger across than a well-grown currant,) and each of these is itself another exquisite feather, with central quill and lateral webs, whose filaments are not to be counted.

The extremity of these breast plumes parts slightly into two, as you see in the peacock's, and many other such decorative ones. The transition from the entirely leaf-like shape of the active plume, with its oblique point, to the more or less symmetrical dualism of the decorative plume, corresponds with the change from the pointed green leaf to the dual, or heart-shaped, petal of many flowers. I shall return to this part of our subject, having given you, I believe, enough of detail for the present.

35. I have said nothing to-day of the mythology of the bird, though I told you that would always be, for us, the most important part of its natural history. But I am obliged, sometimes, to take what we immediately want, rather than what, ultimately, we shall need chiefly. In the second place, you probably, most of you, know more of the mythology of the robin than I do, for the stories about it are all northern, and I know scarcely any myths but the Italian and Greek. You will find

under the name "Robin," in Miss Yonge's exhaustive and admirable "History of Christian Names," the various titles of honor and endearment connected with him, and with the general idea of redness,--from the bishop called "Bright Red Fame," who founded the first great Christian church on the Rhine, (I am afraid of your thinking I mean a pun, in connection with robins, if I tell you the locality of it,) down through the Hoods, and Roys, and Grays, to Robin Goodfellow, and Spenser's "Hobbinol," and our modern "Hob,"--joining on to the "goblin," which comes from the old Greek Kobalos [Greek: Kobalos]. But I cannot let you go without asking you to compare the English and French feeling about small birds, in Chaucer's time, with our own on the same subject. I say English and French, because the original French of the Romance of the Rose shows more affection for birds than even Chaucer's translation, passionate as he is, always, in love for any one of his little winged brothers or sisters. Look, however, either in the French or English at the description of the coming of the God of Love, leading his carol-dance, in the garden of the Rose.

His dress is embroidered with figures of flowers and of beasts; but about him fly the *living* birds. The French is:

Il etoit tout convert d'oisiaulx
De rossignols et de papegaux
De calendre, et de mesangel.
Il semblait que ce fut une angle
Qui fuz tout droit venuz du ciel.

36. There are several points of philology in this transitional French, and in Chaucer's translation, which it is well worth your patience to observe. The monkish Latin "angelus," you see, is passing through the very unpoetical form "angle," into "ange;" but, in order to get a rhyme with it in that angular form, the French troubadour expands the bird's name, "mesange," quite arbitrarily, into "mesangel." Then Chaucer, not liking the "mes" at the beginning of the word, changes that unscrupulously into "arch;" and gathers in, though too shortly, a lovely bit from another place about the nightingales flying so close round Love's head that they strike some of the leaves off his crown of roses; so that the English runs thus:

But nightingales, a full great rout
That flien over his head about,
The leaves felden as they flien
And he was all with birds wrien,
With popinjay, with nightingale,

With chelaundre, and with wodewale,
With finch, with lark, and with archangel.
He seemed as he were an angell,
That down were comen from Heaven clear.

Now, when I first read this bit of Chaucer, without referring to the original, I was greatly delighted to find that there was a bird in his time called an archangel, and set to work, with brightly hopeful industry, to find out what it was. I was a little discomfited by finding that in old botany the word only meant "dead-nettle," but was still sanguine about my bird, till I found the French form descend, as you have seen, into a mesangel, and finally into mesange, which is a provincialism from meion [Greek: meion], and means, the smallest of birds--or, specially here,--a titmouse. I have seldom had a less expected or more ignominious fall from the clouds.

37. The other birds, named here and in the previous description of the garden, are introduced, as far as I can judge, nearly at random, and with no precision of imagination like that of Aristophanes; but with a sweet childish delight in crowding as many birds as possible into the smallest space. The popinjay is always prominent; and I want some of you to help me (for I have not time at present for the chase) in hunting the parrot down on his first appearance in Europe. Just at this particular time he contested favor even with the falcon; and I think it a piece of good fortune that I chanced to draw for you, thinking only of its brilliant color, the popinjay, which Carpaccio allows to be present on the grave occasion of St. George's baptizing the princess and her father.

38. And, indeed, as soon as the Christian poets begin to speak of the singing of the birds, they show themselves in quite a different mood from any that ever occurs to a Greek. Aristophanes, with infinitely more skill, describes, and partly imitates, the singing of the nightingale; but simply as beautiful sound. It "fills the thickets with honey;" and if in the often-quoted--just because it is *not* characteristic of Greek literature--passage of the Coloneus, a deeper sentiment is shown, that feeling is dependent on association of the bird-voices with deeply pathetic circumstances. But this troubadour finds his heart in heaven by the power of the singing only:--

Trop parfoisaient beau servise
Ciz oiselles que je vous devise.
Il chantaient un chant ytel
Com fussent angle esperitel.

We want a moment more of word-chasing to enjoy this. "Oiseau," as you know, comes from "avis;" but it had at this time got "oisel" for its singular number, of which the terminating "sel" confused itself with the "selle," from "ancilla" in domisella and demoiselle; and the feminine form "oiselle" thus snatched for itself some of the delightfulness belonging to the title of a young lady. Then note that "esperitel" does not here mean merely spiritual, (because all angels are spiritual) but an "angle esperitel" is an angel of the air. So that, in English, we could only express the meaning in some such fashion as this:--

They perfected all their service of love,
These maiden birds that I tell you of.
They sang such a song, so finished-fair,
As if they were angels, born of the air.

39. Such were the fancies, then, and the scenes, in which Englishmen took delight in Chaucer's time. England was then a simple country; we boasted, for the best kind of riches, our birds and trees, and our wives and children. We had now grown to be a rich one; and our first pleasure is in shooting our birds; but it has become too expensive for us to keep our trees. Lord Derby, whose crest is the eagle and child--you will find the northern name for it, the bird and bantling, made classical by Scott--is the first to propose that wood-birds should have no more nests. We must cut down all our trees, he says, that we may effectively use the steam-plow; and the effect of the steam-plow, I find by a recent article in the *Cornhill Magazine,* is that an English laborer must not any more have a nest, nor bantlings, neither; but may only expect to get on prosperously in life, if he be perfectly skillful, sober, and honest, and dispenses, at least until he is forty-five, with the "luxury of marriage."

40. Gentlemen, you may perhaps have heard me blamed for making no effort here to teach in the artisans' schools. But I can only say that, since the future life of the English laborer or artisan (summing the benefits to him of recent philosophy and economy) is to be passed in a country without angels and without birds, without prayers and without songs, without trees and without flowers, in a state of exemplary sobriety, and (extending the Catholic celibacy of the clergy into celibacy of the laity) in a state of dispensation with the luxury of marriage, I do not believe he will derive either profit or entertainment from lectures on the Fine Arts.

LECTURE II.[8]

THE SWALLOW.

41. We are to-day to take note of the form of a creature which gives us a singular example of the unity of what artists call beauty, with the fineness of mechanical structure, often mistaken for it. You cannot but have noticed how little, during the years of my past professorship, I have introduced any questions as to the nature of beauty. I avoided them, partly because they are treated of at length in my books; and partly because they are, in the last degree, unpractical. We are born to like or dislike certain aspects of things; nor could I, by any arguments, alter the defined tastes which you received at your birth, and which the surrounding circumstances of life have enforced, without any possibility of your voluntary resistance to them. And the result of those surrounding circumstances, to-day, is that most English youths would have more pleasure in looking at a locomotive than at a swallow; and that many English philosophers would suppose the pleasure so received to be through a new sense of beauty. But the meaning of the word "beauty" in the fine arts, and in classical literature, is properly restricted to those very qualities in which the locomotion of a swallow differs from that of an engine.

42. Not only from that of an engine; but also from that of animals in whose members the mechanism is so complex as to give them a resemblance to engines. The dart of the common house-fly, for instance, in full strength, is a more wonderful movement than that of a swallow. The mechanism of it is not only more minute, but the swiftness of the action so much greater, that the vibration of the wing is invisible. But though a school-boy might prefer the locomotive to the swallow, he would not carry his admiration of finely mechanical velocity into unqualified sympathy with the workmanship of the God of Ekron; and would generally suppose that flies were made only to be food for the more graceful fly-catcher,--whose finer grace you will discover, upon reflection, to be owing to the very moderation and simplicity of its structure, and to the subduing of that infinitude of joints, claws, tissues, veins, and fibers which inconceivably vibrate in the microscopic[9] creature's motion, to a quite intelligible and simple balance of rounded body upon edged plume, maintained not without visible, and sometimes

fatigued, exertion, and raising the lower creature into fellowship with the volition and the virtue of humanity.

43. With the virtue, I say, in an exceedingly qualified sense; meaning rather the strength and art displayed in overcoming difficulties, than any distinct morality of disposition. The bird has kindly and homely qualities; but its principal "virtue" for *us*, is its being an incarnate voracity, and that it moves as a consuming and cleansing power. You sometimes hear it said of a humane person that they would not kill a fly: from 700 to 1,000 flies a day are a moderate allowance for a baby swallow.

44. Perhaps, as I say this, it may occur to some of you to think, for the first time, of the reason of the bird's name. For it is very interesting, as a piece of language study, to consider the different power on our minds,--nay, the different sweetness to the ear,--which, from association, these same two syllables receive, when we read them as a noun, or as a verb. Also, the word is a curious instance of the traps which are continually open for rash etymologists. At first, nothing would appear more natural than that the name should have been given to the bird from its reckless function of devouring. But if you look to your Johnson, you will find, to your better satisfaction, that the name means "bird of porticos," or porches, from the Gothic "swale;" "subdivale,"--so that he goes back in thought as far as Virgil's, "Et nunc porticibus vacuis, nunc humida circum, stagna sonat." Notice, in passing, how a simile of Virgil's, or any other great master's, will probably tell in two or more ways at once. Juturna is compared to the swallow, not merely as winding and turning swiftly in her chariot, but as being a water-nymph by birth,--"Stagnis quae, fluminibusque sonoris, praesidet." How many different creatures in one the swallow is by birth, as a Virgilian simile is many thoughts in one, it would take many more lectures than one to show you clearly; but I will indicate them with such rough sketch as is possible.

45. It belongs, as most of you know, to a family of birds called Fissirostres, or, literally, split-beaks. Split heads would be a better term, for it is the enormous width of mouth and power of gaping which the epithet is meant to express. A dull sermon, for instance, makes half the congregation "fissirostres." The bird, however, is most vigilant when its mouth is widest, for it opens as a net to catch whatever comes in its way,--hence the French, giving the whole family the more literal name, "Gobble-fly"--Gobe-mouche, extend the term to the open-mouthed and too acceptant appearance of a simpleton.

46. Partly in order to provide for this width of mouth, but more for the advantage in flight, the head of the swallow is rounded into a bullet shape, and sunk down on the shoulders, with no neck whatever between, so as to give nearly the aspect of a conical rifle bullet to the entire front of the body; and, indeed, the bird moves more like a bullet than an arrow--dependent on a certain impetus of weight rather than on sharp penetration of the air. I say dependent on, but I have not yet been able to trace distinct relation between the shapes of birds and their powers of flight. I suppose the form of the body is first determined by the general habits and food, and that nature can make any form she chooses volatile; only one point I think is always notable, that a complete master of the art of flight must be short-necked, so that he turns altogether, if he turns at all. You don't expect a swallow to look round a corner before he goes round it; he must take his chance. The main point is that he may be able to stop himself, and turn, in a moment.

47. The stopping, on any terms, is difficult enough to understand; nor less so, the original gaining of the pace. We always think of flight as if the main difficulty of it were only in keeping up in the air;--but the buoyancy is conceivable enough, the far more wonderful matter is the getting along. You find it hard work to row yourself at anything like speed, though your impulse-stroke is given in a heavy element, and your return-stroke in a light one. But both in birds and fishes, the impelling stroke and its return are in the same element; and if, for the bird, that medium yields easily to its impulses, it secedes as easily from the blow that gives it. And if you think what an effort you make to leap six feet, with the earth for a fulcrum, the dart either of a trout or a swallow, with no fulcrum but the water and air they penetrate, will seem to you, I think, greatly marvelous. Yet of the mode in which it is accomplished you will as yet find no undisputed account in any book on natural history, and scarcely, as far as I know, definite notice even of the rate of flight. What do you suppose it is? We are apt to think of the migration of a swallow, as we should ourselves of a serious journey. How long, do you think, it would take him, if he flew uninterruptedly, to get from here to Africa?

48. Michelet gives the rate of his flight (at full speed, of course,) as eighty leagues an hour. I find no more sound authority; but do not doubt his approximate accuracy;[10] still how curious and how provoking it is that neither White of Selborne, Bewick, Yarrell, nor Gould, says a word about this, one should have thought the most interesting, power of the bird.[11]

Taking Michelet's estimate--eighty French leagues, roughly two hundred and fifty miles, an hour--we have a thousand miles in four hours. That is to say, leaving Devonshire after an early breakfast, he could be in Africa to lunch.

49. He could, I say, if his flight were constant; but though there is much inconsistency in the accounts, the sum of testimony seems definite that the swallow is among the most fatiguable of birds. "When the weather is hazy," (I quote Yarrell) "they will alight on fishing-boats a league or two from land, so tired that when any one tries to catch them, they can scarcely fly from one end of the boat to the other."

I have no time to read to you the interesting evidence on this point given by Yarrell, but only that of the brother of White of Selborne, at Gibraltar. "My brother has always found," he himself writes, "that some of his birds, and particularly the swallow kind, are very sparing of their pains in crossing the Mediterranean; for when arrived at Gibraltar, they do not 'set forth their airy caravan, high over seas,' but scout and hurry along in little detached parties of six or seven in a company; and sweeping low, just over the surface of the land and water, direct their course to the opposite continent at the narrowest passage they can find."

50. You will observe, however, that it remains an open question whether this fear of sea may not be, in the swallow, like ours of the desert. The commissariat department is a serious one for birds that eat a thousand flies a day when just out of the egg; and it is possible that the weariness of swallows at sea may depend much more on fasting than flying. Captain (or Admiral?) Sir Charles Wager says that "one spring-time, as he came into soundings in the English Channel, a great flock of swallows came and settled on all his rigging; every rope was covered; they hung on one another like a swarm of bees; even the decks were filled with them. They seemed almost famished and spent, and were only feathers and bone; but, being recruited with a night's rest, took their flight in the morning."

51. Now I detain you on this point somewhat, because it is intimately connected with a more important one. I told you we should learn from the swallow what a wing was. Few other birds approach him in the beauty of it, or apparent power. And yet, after all this care taken about it, he gets tired; and instead of flying, as we should do in his place, all over the world, and tasting the flavor of the midges in every marsh which the infinitude of human folly has left to breed gnats instead of growing corn,--he is of all birds, characteristically, except when he absolutely can't

help it, the stayer at home; and contentedly lodges himself and his family in an old chimney, when he might be flying all over the world.

At least you would think, if he built in an English chimney this year, he would build in a French one next. But no. Michelet prettily says of him, "He is the bird of return." If you will only treat him kindly, year after year, he comes back to the same niche, and to the same hearth, for his nest.

To the same niche; and builds himself an opaque walled house within that. Think of this a little, as if you heard of it for the first time.

52. Suppose you had never seen a swallow; but that its general habit of life had been described to you, and you had been asked, how you thought such a bird would build its nest. A creature, observe, whose life is to be passed in the air; whose beak and throat are shaped with the fineness of a net for the catching of gnats; and whose feet, in the most perfect of the species, are so feeble that it is called the Footless Swallow, and cannot stand a moment on the ground with comfort. Of all land birds, the one that has least to do with the earth; of all, the least disposed, and the least able, to stop to pick anything up. What will it build with? Gossamer, we should say,--thistledown,--anything it can catch floating, like flies.

But it builds with stiff clay.

53. And observe its chosen place for building also. You would think, by its play in the air, that not only of all birds, but of all creatures, it most delighted in space and freedom. You would fancy its notion of the place for a nest would be the openest field it could find; that anything like confinement would be an agony to it; that it would almost expire of horror at the sight of a black hole.

And its favorite home is down a chimney.

54. Not for your hearth's sake, nor for your company's. Do not think it. The bird will love you if you treat it kindly; is as frank and friendly as bird can be; but it does not, more than others, seek your society. It comes to your house because in no wild wood, nor rough rock, can it find a cavity close enough to please it. It comes for the blessedness of imprisonment, and the solemnity of an unbroken and constant shadow, in the tower, or under the eaves.

Do you suppose that this is part of its necessary economy, and that a swallow could not catch flies unless it lived in a hole?

Not so. This instinct is part of its brotherhood with another race of creatures. It is given to complete a mesh in the reticulation of the orders of life.

55. I have already given you several reasons for my wish that you should retain, in classifying birds, the now rejected order of Picae. I am going to read you a passage from Humboldt, which shows you what difficulties one may get into for want of it.

You will find in the second volume of his personal narrative, an account of the cave of Caripe in New Andalusia, which is inhabited by entirely nocturnal birds, having the gaping mouths of the goat-sucker and the swallow, and yet feeding on fruit.

Unless, which Mr. Humboldt does not tell us, they sit under the trees outside, in the night time, and hold their mouths open, for the berries to drop into, there is not the smallest occasion for their having wide mouths, like swallows. Still less is there any need, since they are fruit eaters, for their living in a cavern 1,500 feet out of daylight. They have only, in consequence, the trouble of carrying in the seeds to feed their young, and the floor of the cave is thus covered, by the seeds they let fall, with a growth of unfortunate pale plants, which have never seen day. Nay, they are not even content with the darkness of their cave; but build their nests in the funnels with which the roof of the grotto is pierced like a sieve; live actually in the chimney, not of a house, but of an Egyptian sepulcher! The color of this bird, of so remarkable taste in lodging, Humboldt tells us, is "of dark bluish-gray, mixed with streaks and specks of black. Large white spots, which have the form of a heart, and which are bordered with black, mark the head, the wings, and the tail. The spread of the wings, which are composed of seventeen or eighteen quill feathers, is three feet and a half. Suppressing, with Mr. Cuvier, the order of Picae, we must refer this extraordinary bird to the *Sparrows*."

56. We can only suppose that it must be, to our popular sparrows, what the swallow of the cinnamon country is to our subordinate swallow. Do you recollect the cinnamon swallows of Herodotus, who build their mud-nests in the faces of the cliffs where Dionusos was brought up, and where nobody can get near them; and how the cinnamon merchants fetch them joints of meat, which the unadvised birds, flying up to their nests with, instead of cinnamon,--nest and all come down together,--the original of Sindbad's valley-of-diamond story?

57. Well, Humboldt is reduced, by necessities of recent classification, to call a bird three feet and a half across the wings, a sparrow. I have no right to laugh at him, for I am just going, myself, to call the cheerfulest and brightest of birds of the air, an owl. All these architectural and sepulchral habits, these Egyptian manners of the sand-martin, digging caves in the sand, and border-trooper's habits of the chimney swallow,

living in round towers instead of open air, belonging to them as connected with the tribe of the falcons through the owls! and not only so, but with the mammalia through the bats! A swallow is an emancipated owl, and a glorified bat; but it never forgets its fellowship with night.

58. Its *ancient* fellowship, I had nearly written; so natural is it to think of these similarly-minded creatures, when the feelings that both show are evidently useless to one of them, as if the inferior had changed into the higher. The doctrine of development seems at first to explain all so pleasantly, that the scream of consent with which it has been accepted by men of science, and the shriller vociferation of the public's gregarious applause, scarcely permit you the power of antagonistic reflection. I must justify to-day, in graver tone than usual, the terms in which I have hitherto spoken,--it may have been thought with less than the due respect to my audience,--of the popular theory.

59. Supposing that the octohedrons of galena, of gold, and of oxide of iron, were endowed with powers of reproduction, and perished at appointed dates of dissolution or solution, you would without any doubt have heard it by this time asserted that the octohedric form, which was common to all, indicated their descent from a common progenitor; and it would have been ingeniously explained to you how the angular offspring of this eight-sided ancestor had developed themselves, by force of circumstances, into their distinct metallic perfections; how the galena had become gray and brittle under prolonged subterranean heat, and the gold yellow and ductile, as it was rolled among the pebbles of amber-colored streams.

60. By the denial to these structures of any individually reproductive energy, you are forced to accept the inexplicable (and why expect it to be otherwise than inexplicable?) fact, of the formation of a series of bodies having very similar aspects, qualities, and chemical relations to other substances, which yet have no connection whatever with each other, and are governed, in their relation with their native rocks, by entirely arbitrary laws. It has been the pride of modern chemistry to extricate herself from the vanity of the alchemist, and to admit, with resignation, the independent, though apparently fraternal, natures, of silver, of lead, of platinum,--aluminium,--potassium. Hence, a rational philosophy would deduce the probability that when the arborescence of dead crystallization rose into the radiation of the living tree, and sentient plume, the splendor of nature in her more exalted power would not be restricted to a less variety of design; and the beautiful caprice in which she gave to the silver its frost and to the opal its fire, would not be

subdued under the slow influences of accident and time, when she wreathed the swan with snow, and bathed the dove in iridescence. That the infinitely more exalted powers of life must exercise more intimate influence over matter than the reckless forces of cohesion;--and that the loves and hatreds of the now conscious creatures would modify their forms into parallel beauty and degradation, we might have anticipated by reason, and we ought long since to have known by observation. But this law of its spirit over the substance of the creature involves, necessarily, the indistinctness of its type, and the existence of inferior and of higher conditions, which whole eras of heroism and affection--whole eras of misery and misconduct,--confirm into glory, or confuse into shame. Collecting the causes of changed form, in lower creatures, by distress, or by adaptation,--by the disturbance or intensifying of the parental strength, and the native fortune--the wonder is, not that species should sometimes be confused, but that the greater number of them remain so splendidly, so manifestly, so eternally distinct; and that the vile industries and vicious curiosities of modern science, while they have robbed the fields of England of a thousand living creatures, have not created in them one.

61. But even in the paltry knowledge we have obtained, what unanimity have we?--what security? Suppose any man of ordinary sense, knowing the value of time, and the relative importance of subjects of thought, and that the whole scientific world was agog concerning the origin of species, desired to know first of all--what was meant by a species.

He would naturally look for the definition of species first among the higher animals, and expect it to be best defined in those which were best known. And being referred for satisfaction to the 226th page of the first volume of Mr. Darwin's "Descent of Man," he would find this passage:--

"Man has been studied more carefully than any other organic being, and yet there is the greatest possible diversity among capable judges, whether he should be classed as a single species or race, or as two (Virey), as three (Jacquinot), as four (Kant), five (Blumenbach), six (Buffon), seven (Hunter), eight (Agassiz), eleven (Pickering), fifteen (Bory St. Vincent), sixteen (Desmoulins), twenty-two (Morton), sixty (Crawford), or as sixty-three according to Burke."

And in the meantime, while your men of science are thus vacillating, in the definition of the species of the only animal they have the opportunity of studying inside and out, between one and sixty-three; and disputing about the origin, in past ages, of what they cannot define in the present ones; and deciphering the filthy heraldries which record the relation of humanity to the ascidian and the crocodile, you have ceased utterly to distinguish between the two species of man, evermore separate by infinite separation: of whom the one, capable of loyalty and of love, can at least conceive spiritual natures which have no taint from their own, and leave behind them, diffused among thousands on earth, the happiness they never hoped, for themselves, in the skies; and the other, capable only of avarice, hatred, and shame, who in their lives are the companions of the swine, and leave in death nothing but food for the worm and the vulture.

62. Now I have first traced for you the relations of the creature we are examining to those beneath it and above, to the bat and to the falcon. But you will find that it has still others to entirely another world. As you watch it glance and skim over the surface of the waters, has it never struck you what relation it bears to the creatures that glance and glide *under* their surface? Fly-catchers, some of them, also,--fly-catchers in the same manner, with wide mouth; while in motion the bird almost exactly combines the dart of the trout with the dash of the dolphin, to the rounded forehead and projecting muzzle of which its own bullet head and bill exactly correspond. In its plunge, if you watch it bathing, you may see it dip its breast just as much under the water as a porpoise shows its back above. You can only rightly describe the bird by the resemblances, and images of what it seems to have changed from,--then adding the fantastic and beautiful contrast of the unimaginable change. It is an owl that has been trained by the Graces. It is a bat that loves the morning light. It is the aerial reflection of a dolphin. It is the tender domestication of a trout.

63. And yet be assured, as it cannot have been all these creatures, so it has never, in truth, been any of them. The transformations believed in by the mythologists are at least spiritually true; you cannot too carefully trace or too accurately consider them. But the transformations believed in by the anatomist are as yet proved true in no single instance, and in no substance, spiritual or material; and I cannot too often, or too earnestly, urge you not to waste your time in guessing what animals may once have been, while you remain in nearly total ignorance of what they are.

64. Do you even know distinctly from each other,--(for that is the real naturalist's business; instead of confounding them with each other),--do you know distinctly the five great species of this familiar bird?--the swallow, the house-martin, the sand-martin, the swift, and the Alpine swift?--or can you so much as answer the first question which would

suggest itself to any careful observer of the form of its most familiar species,--yet which I do not find proposed, far less answered, in any scientific book,--namely, why a swallow has a swallow-tail?

It is true that the tail feathers in many birds appear to be entirely,--even cumbrously, decorative; as in the peacock, and birds of paradise. But I am confident that it is not so in the swallow, and that the forked tail, so defined in form and strong in plume, has indeed important functions in guiding the flight; yet notice how surrounded one is on all sides with pitfalls for the theorists. The forked tail reminds you at once of a fish's; and yet, the action of the two creatures is wholly contrary. A fish lashes himself forward with his tail, and steers with his fins; a swallow lashes himself forward with his fins, and steers with his tail; partly, not necessarily, because in the most dashing of the swallows, the swift, the fork of the tail is the least developed. And I never watch the bird for a moment without finding myself in some fresh puzzle out of which there is no clue in the scientific books. I want to know, for instance, how the bird turns. What does it do with one wing, what with the other? Fancy the pace that has to be stopped; the force of bridle-hand put out in an instant. Fancy how the wings must bend with the strain; what need there must be for the perfect aid and work of every feather in them. There is a problem for you, students of mechanics,--How does a swallow turn?

You shall see, at all events, to begin with, to-day, how it gets along.

65. I say you shall see; but indeed you have often seen, and felt,--at least with your hands, if not with your shoulders,--when you chanced to be holding the sheet of a sail.

I have said that I never got into scrapes by blaming people wrongly; but I often do by praising them wrongly. I never praised, without qualification, but one scientific book in my life (that I remember)--this of Dr. Pettigrew's on the Wing;[12] and now I must qualify my praise considerably, discovering, when I examined the book farther, that the good doctor had described the motion of a bird as resembling that of a kite, without ever inquiring what, in a bird, represented that somewhat important part of a kite, the string. You will, however, find the book full of important observations, and illustrated by valuable drawings. But the point in question you must settle for yourselves, and you easily may. Some of you perhaps, knew, in your time, better than the doctor, how a kite stopped; but I do not doubt that a great many of you also know, now, what is much more to the purpose, how a ship gets along. I will take the simplest, the most natural, the most beautiful of sails,--the lateen sail of the Mediterranean.

66. I draw it rudely in outline, as it would be set for a side-wind on the boat you probably know best,--the boat of burden on the Lake of Geneva (Fig. 3), not confusing the drawing by adding the mast, which, you know, rakes a little, carrying the yard across it (*a*). Then, with your permission, I will load my boat thus, with a few casks of Vevay vintage--and, to keep them cool, we will put an awning over them, so (*b*). Next, as we are classical scholars, instead of this rustic stern of the boat, meant only to run easily on a flat shore, we will give it an Attic embolon [Greek: embolon] (*c*). (We have no business, indeed, yet, to put an embolon [Greek: embolon] on a boat of burden, but I hope some day to see all our ships of war loaded with bread and wine, instead of artillery.) Then I shade the entire form (*c*); and, lastly, reflect it in the water (*d*)--and you have seen something like that before, besides a boat, haven't you?

FIG. 3.

There is the gist of the whole business for you, put in very small space; with these only differences: in a boat, the air strikes the sail; in a bird, the sail strikes the air: in a boat, the force is lateral, and in a bird downwards; and it has its sail on both sides. I shall leave you to follow out the mechanical problem for yourselves, as far as the mere resolution of force is concerned. My business, as a painter, is only with the exquisite organic weapon that deals with it.

67. Of which you are now to note farther, that a bird is required to manage his wing so as to obtain two results with one blow:--he has to keep himself up, as well as to get along.

But observe, he only requires to keep himself up *because* he has to get along. The buoyancy might have been given at once, if nature had wanted *that* only; she might have blown the feathers up with the hot air of the breath, till the bird rose in air like a cork in water. But it has to be, not a buoyant cork, but a buoyant *bullet*. And therefore that it may have momentum for pace, it must have weight to carry; and to carry that weight, the wings must deliver their blow with effective vertical, as well as oblique, force.

Here, again, you may take the matter in brief sum. Whatever is the ship's loss, is the bird's gain; whatever tendency the ship has to leeway, is all given to the bird's support, so that every atom[13] of force in the blow is of service.

68. Therefore you have to construct your organic weapon, so that this absolutely and perfectly economized force may be distributed as the bird chooses at any moment. That, if it wants to rise, it may be able to strike vertically more than obliquely;--if the order is, go-ahead, that it may put the oblique screw on. If it wants to stop in an instant, that it may be able to throw its wings up full to the wind; if it wants to hover, that it may be able to lay itself quietly on the wind with its wings and tail, or, in calm air, to regulate their vibration and expansion into tranquillity of gliding, or of pausing power. Given the various proportions of weight and wing; the conditions of possible increase of muscular force and quill-strength in proportion to size; and the different objects and circumstances of flight,--you have a series of exquisitely complex problems, and exquisitely perfect solutions, which the life of the youngest among you cannot be long enough to read through so much as once, and of which the future infinitudes of human life, however granted or extended, never will be fatigued in admiration.

69. I take the rude outline of sail in Fig. 3, and now considering it as a jib of one of our own sailing vessels, slightly exaggerate the loops at the edge, and draw curved lines from them to the opposite point, Fig. 4; and I have a reptilian or dragon's wing, which would, with some ramification of the supporting ribs, become a bat's or moth's; that is to say, an extension of membrane between the ribs (as in an umbrella), which will catch the wind, and flutter upon it, like a leaf; but cannot strike it to any purpose. The flying squirrel drifts like a falling leaf; the bat flits like a black rag torn at the edge. To give power, we must have plumes that can strike, as with the flat of a sword-blade; and to give *perfect* power, these must be laid over each other, so that each may support the one below it. I use the word below advisedly: we have to strike *down*. The lowest feather is the one that first meets the adverse force. It is the one to be supported.

FIG. 4.

Now for the manner of the support. You must all know well the look of the machicolated parapets in mediaeval castles. You know they are carried on rows of small projecting buttresses constructed so that, though the uppermost stone, far-projecting, would break easily under any shock, it is supported by the next below, and so on, down to the wall. Now in this figure I am obliged to separate the feathers by white spaces, to show you them distinctly. In reality they are set as close to each other as can be, but putting them as close as I can, you get *a* or *b*, Fig. 5, for the rough section of the wing, thick towards the bird's head, and curved like a sickle, so that in striking down it catches the air, like a reaping-hook, and in rising up, it throws off the air like a pent-house.

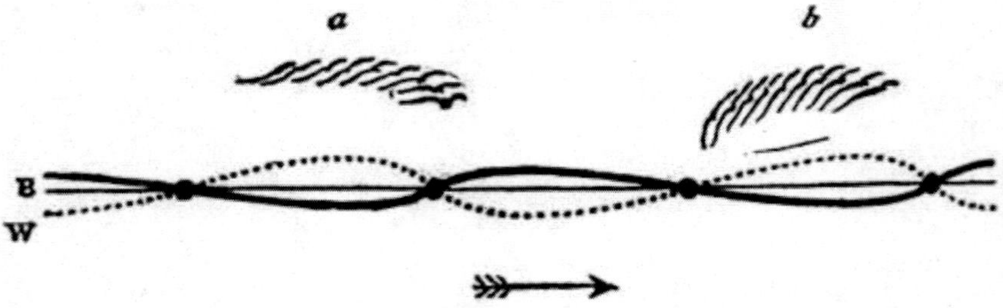

FIG. 5.

70. The stroke would therefore be vigorous, and the recovery almost effortless, were even the direction of both actually vertical. But they are vertical only with relation to the bird's body. In space they follow the forward flight, in a softly curved line; the downward stroke being as effective as the bird chooses, the recovery scarcely encounters resistance in the softly gliding ascent. Thus, in Fig. 5, (I can only explain this to readers a little versed in the elements of mechanics,) if B is the locus of the center of gravity of the bird, moving in slow flight in the direction of the arrow, w is the locus of the leading feather of its wing, and *a* and *b*, roughly, the successive positions of the wing in the down-stroke and recovery.

71. I say the down-stroke is as effective as the bird chooses; that is to say, it can be given with exactly the quantity of impulse, and exactly the quantity of supporting power, required at the moment. Thus, when the bird wants to fly slowly, the wings are fluttered fast, giving vertical blows; if it wants to pause absolutely in still air, (this large birds cannot do, not being able to move their wings fast enough,) the velocity becomes vibration, as in the humming-bird: but if there is wind, any of the larger birds can lay themselves on it like a kite, their own weight answering the purpose of the string,[14] while they keep the wings and tail in an inclined plane, giving them as much gliding ascent as counteracts the fall. They nearly all, however, use some slightly gliding force at the same time; a single stroke of the wing, with forward intent, seeming enough to enable them to glide on for half a minute or more without stirring a plume. A circling eagle floats an inconceivable time without visible stroke: (fancy the pretty action of the inner wing, *backing* air instead of water, which gives exactly the breadth of circle he chooses). But for exhibition of the complete art of flight, a swallow on rough water is the master of masters. A sea-gull, with all its splendid power, generally has its work cut out for it, and is visibly fighting; but the

swallow plays with wind and wave as a girl plays with her fan, and there are no words to say how many things it does with its wings in any ten seconds, and does consummately. The mystery of its dart remains always inexplicable to me; no eye can trace the bending of bow that sends that living arrow.

But the main structure of the noble weapon we may with little pains understand.

72. In the sections *a* and *b* of Fig. 5, I have only represented the quills of the outer part of the wing. The relation of these, and of the inner quills, to the bird's body may be very simply shown.

Fig. 6 is a rude sketch, typically representing the wing of any bird, but actually founded chiefly on the sea-gull's.

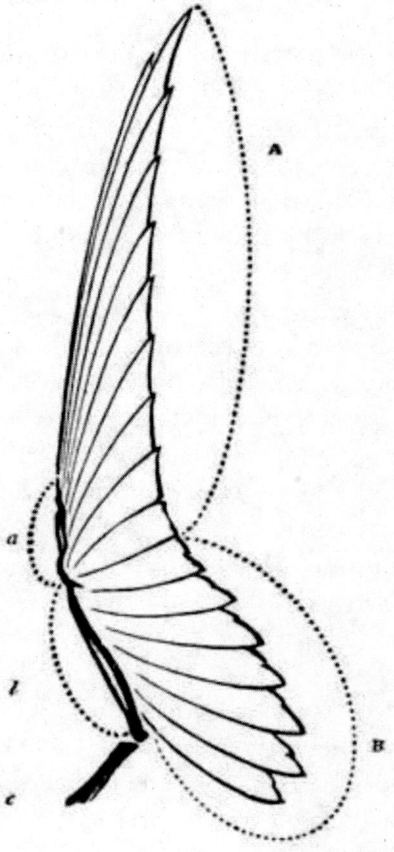

FIG. 6.

It is broadly composed of two fans, A and B. The out-most fan, A, is carried by the bird's hand; of which I rudely sketch the contour of the bones at *a*. The innermost fan, B, is carried by the bird's forearm, from wrist to elbow, *b*.

The strong humerus, *c*, corresponding to our arm from shoulder to elbow, has command of the whole instrument. No feathers are attached to this bone; but covering and protecting ones are set in the skin of it, completely filling, when the active wing is open, the space between it and the body. But the plumes of the two great fans, A and B, are set into the bones; in Fig. 8, farther on, are shown the projecting knobs on the main arm bone, set for the reception of the quills, which make it look like the club of Hercules. The connection of the still more powerful quills of the outer fan with the bones of the hand is quite beyond all my poor anatomical perceptions, and, happily for me, also beyond needs of artistic investigation.

73. The feathers of the fan A are called the primaries. Those of the fan B, secondaries. Effective actions of flight, whether for support or forward motion, are, I believe, all executed with the primaries, every one of which may be briefly described as the strongest cimeter that can be made of quill substance; flexible within limits, and elastic at its edges--carried by an elastic central shaft--twisted like a windmill sail--striking with the flat, and recovering with the edge.

The secondary feathers are more rounded at the ends, and frequently notched; their curvature is reversed to that of the primaries; they are arranged, when expanded, somewhat in the shape of a shallow cup, with the hollow of it downwards, holding the air therefore, and aiding in all the pause and buoyancy of flight, but little in the activity of it. Essentially they are the brooding and covering feathers of the wing; exquisitely beautiful--as far as I have yet seen, *most* beautiful--in the bird whose brooding is of most use to us; and which has become the image of all tenderness. "How often would I have gathered thy children ... and ye would not."

74. Over these two chief masses of the plume are set others which partly complete their power, partly adorn and protect them; but of these I can take no notice at present. All that I want you to understand is the action of the two main masses, as the wing is opened and closed.

Fig. 7 roughly represents the upper surface of the main feathers of the wing closed. The secondaries are folded over the primaries; and the primaries shut up close, with their outer edges parallel, or nearly so. Fig. 8 roughly shows the outline of the bones, in this position, of one of the larger pigeons.[15]

FIG. 7.

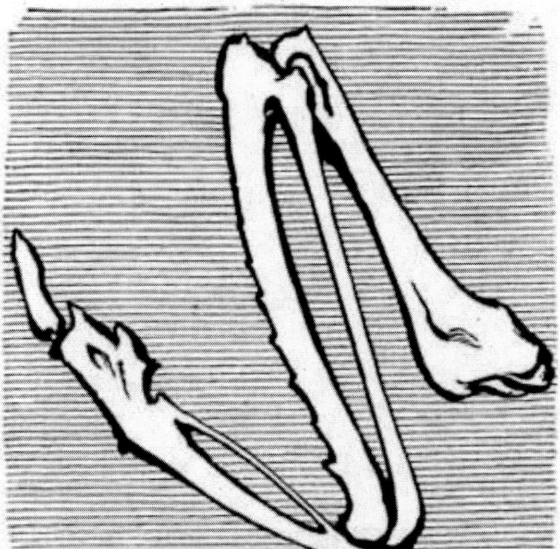

FIG. 8.

75. Then Fig. 9 is (always sketched in the roughest way) the outer, Fig. 10 the inner, surface of a sea-gull's wing in this position. Next, Fig. 11 shows the tops of the four lowest feathers in Fig. 9, in mere outline; A separate (pulled off, so that they can be set side by side), B shut up close in the folded wing, C, opened in the spread wing.

FIG. 9.

FIG. 10.

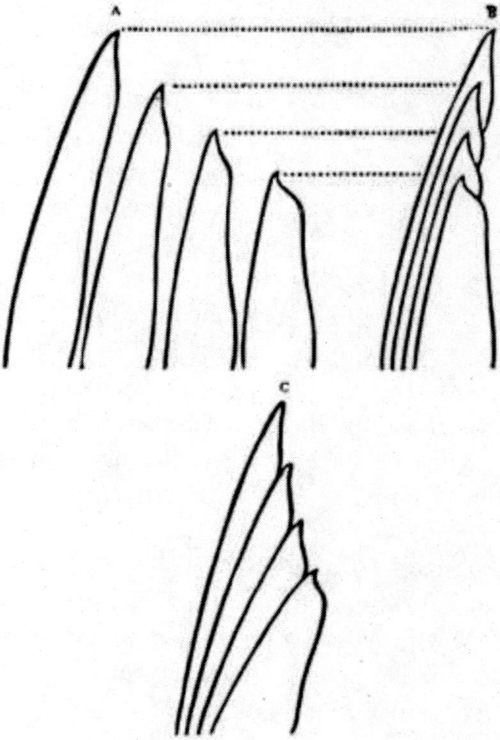

FIG. 11.

76. And now, if you will yourselves watch a few birds in flight, or opening and closing their wings to prune them, you will soon know as much as is needful for our art purposes; and, which is far more desirable, feel how very little we know, to any purpose, of even the familiar creatures that are our companions.

Even what we have seen to-day[16] is more than appears to have been noticed by the most careful painters of the great schools; and you will continually fancy that I am inconsistent with myself in pressing you to learn, better than they, the anatomy of birds, while I violently and constantly urge you to refuse the knowledge of the anatomy of men. But you will find, as my system develops itself, that it is absolutely consistent throughout. I don't mean, by telling you not to study human anatomy, that you are not to know how many fingers and toes you have, nor how you can grasp and walk with them; and, similarly, when you look at a bird, I wish you to know how many claws and wing-feathers it has, and how it grips and flies with them. Of the bones, in either, I shall show you little; and of the muscles, nothing but what can be seen in the living creature, nor, often, even so much.

77. And accordingly, when I now show you this sketch of my favorite Holbein, and tell you that it is entirely disgraceful he should not know what a wing was, better, I don't mean that it is disgraceful he should not know the anatomy of it, but that he should never have looked at it to see how the feathers lie.

Now Holbein paints men gloriously, but never looks at birds; Gibbons, the wood-cutter, carves birds, but can't men;--of the two faults the last is the worst; but the right is in looking at the whole of nature in due comparison, and with universal candor and tenderness.

78. At the whole of nature, I say, not at *super*-nature--at what you suppose to be above the visible nature about you. If you are not inclined to look at the wings of birds, which God has given you to handle and to see, much less are you to contemplate, or draw imaginations of, the wings of angels, which you can't see. Know your own world first--not denying any other, but being quite sure that the place in which you are now put is the place with which you are now concerned; and that it will be wiser in you to think the gods themselves may appear in the form of a dove, or a swallow, than that, by false theft from the form of dove or swallow, you can represent the aspect of gods.

79. One sweet instance of such simple conception, in the end of the Odyssey, must surely recur to your minds in connection with our subject of to-day, but you may not have noticed the recurrent manner in which Homer insists on the thought. When Ulysses first bends and strings his bow, the vibration of the chord is shrill, "like the note of a swallow." A poor and unwarlike simile, it seems! But in the next book, when Ulysses stands with his bow lifted, and Telemachus has brought the lances, and laid them at his feet, and Athena comes to his side to encourage him,--do you recollect the gist of her speech? "You fought," she says, "nine years for the sake of Helen, and for another's house:--now, returned, after all those wanderings, and under your own roof, for it, and its treasures, will you not fight, then?" And she herself flies up to the house-roof, and thence, *in the form of the swallow*, guides the arrows of vengeance for the violation of the sanctities of home.

80. To-day, then, I believe verily for the first time, I have been able to put before you some means of guidance to understand the beauty of the bird which lives with you in your own houses, and which purifies for you, from its insect pestilence, the air that you breathe. Thus the sweet domestic thing has done, for men, at least these four thousand years. She

has been their companion, not of the home merely, but of the hearth, and the threshold; companion only endeared by departure, and showing better her loving-kindness by her faithful return. Type sometimes of the stranger, she has softened us to hospitality; type always of the suppliant, she has enchanted us to mercy; and in her feeble presence, the cowardice, or the wrath, of sacrilege has changed into the fidelities of sanctuary. Herald of our summer, she glances through our days of gladness; numberer of our years, she would teach us to apply our hearts to wisdom;--and yet, so little have we regarded her, that this very day, scarcely able to gather from all I can find told of her enough to explain so much as the unfolding of her wings, I can tell you nothing of her life--nothing of her journeying: I cannot learn how she builds, nor how she chooses the place of her wandering, nor how she traces the path of her return. Remaining thus blind and careless to the true ministries of the humble creature whom God has really sent to serve us, we in our pride, thinking ourselves surrounded by the pursuivants of the sky, can yet only invest them with majesty by giving them the calm of the bird's motion, and shade of the bird's plume:--and after all, it is well for us, if, when even for God's best mercies, and in His temples marble-built, we think that, "with angels and archangels, and all the company of Heaven, we laud and magnify His glorious name"--well for us, if our attempt be not only an insult, and His ears open rather to the inarticulate and unintended praise, of "the Swallow, twittering from her straw-built shed."

LECTURE III.

THE DABCHICKS.

81. I believe that somewhere I have already observed, but permit myself, for immediate use, to repeat what I cannot but think the sagacious observation,--that the arrangement of any sort of animals must be, to say the least, imperfect, if it be founded only on the characters of their feet. And, of all creatures, one would think birds were those which, continually dispensing with the use of their feet, would require for their classification some attention also to be paid to their bodies and wings,--not to say their heads and tails. Nevertheless, the ornithological arrangement at present in vogue may suffice for most scientific persons; but in grouping birds, so that the groups may be understood and remembered by children, I must try to make them a little more generally descriptive.

82. In talking of parrots, for instance, it is only a small part of the creature's nature which is told by its scientific name of 'Scansor,' or 'Climber.' That it only clutches with its claws, and does not snatch or strike with them;--that it helps itself about with its beak, on branches, or bars of cage, in an absurd manner, as if partly imagining itself hung up in a larder, are by no means the most vital matters about the bird. Whereas, that its beak is always extremely short, and is bent down so roundly that the angriest parrot cannot peck, but only *bite*, if you give it a chance; that it *can* bite, pinch, or otherwise apply the mechanism of a pair of nut-crackers from the back of its head, with effect; that it has a little black tongue capable of much talk; above all, that it is mostly gay in plumage, often to vulgarity, and always to pertness;--all these characters should surely be represented to the apprehensive juvenile mind, in sum; and not merely the bird's climbing qualities.

83. Again, that the race of birds called in Latin 'Rasores' *do*, in the search for their food, usually scratch, and kick out their legs behind, living for the most part in gravelly or littery places, of which the hidden treasures are only to be discovered in that manner, seems to me no supremely interesting custom of the animal's life, but only a *manner* of its household, or threshold, economy. But that the tribe, on the whole, is unambitiously domestic, and never predatory; that they fly little and low, eat much of what they can pick up without trouble--and are *themselves* always excellent eating;--yet so exemplary in their own

domestic cares and courtesies that one is ashamed to eat them except in eggs;--that their plumage is for the most part warm brown, delicately and even bewitchingly spotty;--and that, in the goodliest species, the spots become variegated, and inlaid as in a Byzantine pavement, deepening to imperial purple and azure, and lightening into luster of innumerable eyes;--all this, I hold, very clearly and positively, should be explained to children as a part of science, quite as exact, and infinitely more gracious, than that which reckons up the whole tribe of loving and luminous creatures under the feebly descriptive term of 'Scratchers.'

I will venture therefore to recommend my younger readers, in classing birds, to think of them literally from top to toe--from toe to top I should say,--foot, body, and head, studying, with the body, the wings that bear it; and with the head, what brains it can bring to bear on practical matters, and what sense on sentimental. But indeed, primarily, you have to consider whether the bird altogether may not be little more than a fat, cheerful little stomach, in a spotted waistcoat, and with legs to it. That is the main definition of a great many birds--meant to eat all day, chiefly, grubs, or grain--not at all, unless under wintry and calamitous conditions, meant to fast painfully, or be in concern about their food. Faultless in digestion--dinner lasting all day long, with the delight of social intercourse--various chirp and chatter. Flying or fluttering in a practical, not stately, manner: hopping and creeping intelligently. Sociable to man extremely, building and nestling and rustling about him,--prying and speculating, curiously watchful of him at his work, if likely to be profitable to themselves, or even sometimes in mere pitying sympathy, and wonder how such a wingless and beakless creature can do *any*thing.[17]

84. The balance of this kind of bird on its legs is a very important part of its--diagnosis; (we must have a fine word now and then!) Its action on the wing, is mere flutter or flirt, in and out of the hedge, or over it; but its manner of perch, or literally 'bien-seance,' is admirable matter of interest. So also in the birds which are on the water what these are on land; picking up anything anywhere; lazy and fortunate, mostly, themselves; fat, floating, daintiest darlings;--*their* balance on the water, also, and under it, in 'ducking,' a most essential part of their business and being.

85. Then, directly opposed to these, in both kinds, you have the birds which must fast long, and fly far, and watch or fight for their food. Not stomachic in profile; far from cheerful in disposition; more or less lonely in habit; or, if gregarious, out of the way of men. The balance of these on the wing, is no less essential a part of their picturing, than that of the

buntings, robins, and ducks on the foot, or breast: and therefore, especially the position of the head in flying.

86. Accordingly, for complete ornithology, *every* bird must be drawn, as every flower for good botany, both in profile, and looking down upon it: but for the perchers, the standing profile is the most essential; and for the falcons and gulls, the flying *plan*,--the outline of the bird, as it would be seen looking down on it, when its wings were full-spread.

Then, in connection with these general outlines, we want systematic plan and profile of the foot and head; but since we can't have everything at once, let us say the plan of the foot, and profile of the head, quite accurately given; and for every bird consistently, and to scale.

Profile and plan in outline; then, at least the *head* in light and shade, from life, so as to give the expression of the eye. Fallacious, this latter, often, as an indication of character; but deeply significant of habit and power: thus the projecting, full, bead, which enables the smaller birds to see the smallest insect or grain with good in it, gives them much of their bright and often arch expression; while the flattened iris under the beetling brow of the falcons,--projecting, not in frown, but as roof, to shade the eye from interfering skylight,--gives them their apparently threatening and ominous gaze; the iris itself often wide and pale, showing as a lurid saturnine ring under the shadow of the brow plumes.

87. I speak of things that are to be: very assuredly they will be done, some day--not far off, by painters educated as gentlemen, in the strictest sense--working for love and truth, and not for lust and gold. Much has already been done by good and earnest draughtsmen, who yet had not received the higher painter's education, which would have enabled them to see the bird in the greater lights and laws of its form. It is only here and there, by Durer, Holbein, Carpaccio, or other such men, that we get a living bird rightly drawn;[18] but we may be greatly thankful for the unspared labor, and attentive skill, with which many illustrations of ornithology have been produced within the last seventy or eighty years. Far beyond rivalship among them, stands Le Vaillant's monograph, or dualgraph, on the Birds of Paradise, and Jays: its plates, exquisitely engraved, and colored with unwearying care by hand, are insuperable in plume-texture, hue, and action,--spoiled in effect, unhappily, by the vulgar boughs for sustentation. Next, ranks the recently issued history of the birds of Lombardy; the lithographs by Herr Oscar Dressler, superb, but the coloring (chromo-lithotint) poor: and then, the self-taught, but in some qualities greatly to be respected, art of Mr. Gould. Of which, I would fain have spoken with gratitude and admiration in his lifetime;

had not I known, that the qualified expressions necessary for true estimate of his published plates, would have caused him more pain, than any general praise could have counteracted or soothed. Without special criticism, and rejoicing in all the pleasure which any of my young pupils may take in his drawing,--only guarding them, once for all, against the error of supposing it exemplary as art,--I use his plates henceforward for general reference; finding also that, following Mr. Gould's practical and natural arrangement, I can at once throw together in groups, easily comprehensible by British children, all they are ever likely to see of British or Britain-visitant birds: which I find fall, with frank casting, into these following divisions, not in any important matters varying from the usual ones, and therefore less offensive, I hope, to the normal zoologist than my heresies in botany; while yet they enable me to make what I have to say about our native birds more simply presentable to young minds.[19]

88. 1. The HAWKS come first, of course, massed under the single Latin term 'Falco,' and next them,

2. The OWLS second, also of course,--unmistakable, these two tribes, in all types of form, and ways of living.

3. The SWALLOWS I put next these, being connected with the owls by the Goatsucker, and with the falcons by their flight.

4. The PIES next, whose name has a curious double meaning, derived partly from the notion of their being painted or speckled birds; and partly from their being, beyond all others, pecking, or pickax-beaked, birds. They include, therefore, the Crows, Jays, and Woodpeckers; historically and practically a most important order of creatures to man. Next which, I take the great company of the smaller birds of the dry land, under these following more arbitrary heads.

5. The SONGSTERS. The Thrush, Lark, Blackbird, and Nightingale, and one or two choristers more. These are connected with the pheasants in their speckledness, and with the pies in pecking; while the nightingale leads down to the smaller groups of familiar birds.

6. The ROBINS, going on into the minor warblers, and the Wrens; the essential character of a Robin being that it should have some front red in its dress somewhere; and the Cross-bills being included in the class, partly because they have red in their dress, and partly because I don't know where else to put them.

7. The CREEPERS and TITS--separated chiefly on the ground of their minuteness, and subtle little tricks and graces of movement.

8. The SPARROWS, going on into Buntings and Finches.

9. The PHEASANTS (substituting this specific name for that of Scratchers).

10. The HERONS; for the most part wading and fishing creatures, but leading up to the Stork, and including any long-legged birds that run well, such as the Plovers.

11. The DABCHICKS--the subject of our present chapter.

12. The SWANS and GEESE.

13. The DUCKS.

14. The GULLS.

Of these, I take the Dabchicks first, for three sufficient reasons;--that they give us least trouble,--that they best show what I mean by broad principles of grouping,--and that they are the effective clasp, if not center, of all the series; since they are the true link between land and water birds. We will look at one or two of their leading examples, before saying more of their position in bird-society. I shall give for the heading of each article, the name which I propose for the bird in English children's schools--*Dame*-schools if possible; a perfectly simple Latin one, and a familiar English one. The varieties of existing nomenclature will be given in the Appendix, so far as I think them necessary to be known or remembered.

I.

MERULA FONTIUM. TORRENT-OUZEL.

89. There are very few good popular words which do not unite two or more ideas, being founded on one, and catching up others as they go along. Thus I find 'dabchick' to be a corruption of 'dipchick,' meaning birds that only dip, and do not dive, or even duck, for any length of time: but in its broader and customary use it takes up the idea of dabbling; and, as a class-name, stands for 'dabbling-chick,' meaning a bird of small size, that neither wades, nor dives, nor runs, nor swims, nor flies, in a consistent manner; but humorously dabbles, or dips, or flutters, or trips, or plashes, or paddles, and is always doing all manner of odd and delightful things: being also very good-humored, and in consequence, though graceful, inclined to plumpness;[20] and though it never waddles, sometimes, for a minute or two, 'toddles,' and now and then looks more like a ball than a bird. For the most part, being clever, they are also brave, and would be as tame as any other chickens, if we would let them. They are mostly shore birds, living at the edge of irregularly broken water, either streams or sea; and the representative of the whole group with which we will begin is the mysterious little water-ouzel, or

'oiselle,' properly the water-blackbird,--Buffon's 'merle d'eau'--for ouzel is the classic and poetic word for the blackbird, or ouzel-*cock*, "so black of hue," in 'Midsummer Night's Dream.' Johnson gives it from the Saxon 'osle'; but in Chaucer it must be understood simply as the feminine of oiseau. The bird in question might, however, be more properly called, as Bewick calls it, 'water pyot,' or water magpie, for only its back and wings are black,--its head brown, and breast snow white.

90. And now I must, once for all, get over a difficulty in the description of birds' costume. I can always describe the neck-feathers, as such, when birds have any neck to speak of; but when, as the majority of dabchicks, they have not any,--instead of talking of 'throat-feathers' and 'stomach-feathers,' which both seem to me rather ugly words, I shall call the breast feathers the 'chemisette,' and all below them the 'bodice.'

I am now able, without incivility, to distinguish the two families of Water-ouzel. Both have white chemisettes, but the common water-ouzel (Cinclus aquaticus of Gould) has a white bodice, and the other a black one, the bird being called therefore, in ugly Greek, 'Melanogaster,' 'black-stomached.' The black bodice is Norwegian fashion--the white, English; and I find that in Switzerland there is an intermediate Robin-ouzel, with a red bodice: but the ornithologists are at variance as to his 'specific' existence. The chemisette is always white.

91. However dressed, and wherever born, the Ouzel is essentially a mountain-torrent bird, and, Bewick says, may be seen perched on a stone in the midst of a stream, in a continual *dipping* motion, or short curtsey often repeated, while it is watching for its food, which consists of small fishes and insects,--water insects, that is to say, caught mostly at the bottom; many-legged and shrimpy things, according to Gould's plate. The popular tradition that it can walk under the water has been denied by scientific people; but there is no doubt whatever of the fact,--see the authentic evidence of it in the delightful little monograph of the bird published by the Carlisle Naturalist's Society; but how the thing is done nobody but the ouzel knows. Its strong little feet, indeed, have plenty of grip in them, but cannot lay hold of smooth stones, and Mr. Gould himself does not solve the problem. "Some assert that it is done by clinging to the pebbles with its strong claws; others, by considerable exertion and a rapid movement of the wings. Its silky plumage is impervious to wet; and hence when the bird returns to the surface, the pearly drops which roll off into the stream are the only evidence of its recent submersion. It is, indeed, very interesting to observe *this pretty bird walk down a stone, quietly descend into the water*, rise again perhaps at a

distance of several yards down the stream, and 'fly'[21] back to the place it had just left, to perform the same maneuver the next minute, the silence of the interval broken by its cheerful warbling song."

92. In which, you see, we have the reason for its being called 'water-blackbird,' being, I think, the only one of the dabchicks that really sings. Some of the others, (sandpipers) pipe; and others, the stints, say 'stint' in a charming manner; but none of them *sing* except the oiselle. Very singularly, the black-bodiced one seems to like living near manufactories. "The specimen in the Norwich Museum," says Mr. Gould, "is the one mentioned by Mr. Lubbock, in 1845, as 'lately' shot at Hellesdon Mills; and two others are stated by the same author to have been seen at different times by trustworthy observers at Marlingford and Saxthorpe. Of more recent occurrence I may mention a male in my own collection, which was brought to me in the flesh, having been shot in November, 1855, whilst hovering over the river between the foundry bridge and the ferry. It is not a little singular that a bird so accustomed to the clear running streams of the north, and the quiet haunts of the 'silent angler,' should be found, as in this case, almost within the walls of the city, sporting over a river turbid and discolored from the neighboring factories, and with the busy noise of traffic on every side. About the same time that this bird appeared near the city, three others were observed on more than one occasion on the Earlham river, by Mr. Fountaine, of Easton, who is well acquainted with our British birds; but these suddenly disappeared, and were not seen again."

And all will disappear, and never be seen again, but in skeleton, ill-covered with camphorated rags of skin, under the present scientific dispensation; unless some kind-hearted northern squire will let them have the run and the dip of his brooks; and teach the village children to let them alone if they like to wade down to the village.

I am sixty-two, and have passed as much time out of those years by torrent sides as most people. But I have never seen a water-ouzel alive.

II.

ALLEGRETTA NYMPHAEA. LILY-OUZEL.

93. We have got so far, by help of our first example, in the etymology of our entire class, as to rest in the easily memorable root 'dab,' short for dabble, as the foundation of comprehensive nomenclature. But the earlier (if not Aryan!) root 'dip,' must be taken good heed to, also, because, as we further study the customs of aquatic chickens, we shall find that they really mass themselves under the three great heads of

'Duckers,' birds that duck their heads only, and stick up their tails in the air;--'Dippers,' birds that take real dips under, but not far down, in shallow water mostly, for things at the bottom, or else to get out of harm's way, staying down about as long as we could ourselves, if we were used to it;--and 'Divers,' who plunge like stones when they choose,--can go nobody knows how deep in the deep sea,--and swim under the water just as comfortably as upon it, and as fast, if not faster.

But although this is clearly the practical and poetical division, we can't make it a scientific one; for the dippers and dabblers are so like each other that we must take them together; and so also the duckers and divers are inseparable in some of their forms: so that, for convenience of classing, we must keep to the still more general rank I have given--dabchick, duck, and gull,--the last being essentially the aerial sea-bird, which *lives* on the wing.

94. But there is yet one more 'mode of motion' to be thought of, in the class we are now examining. Several of them ought really to be described, not as dipchicks, but as *trip*-chicks; being, as far as I can make out, little in the habit of going under water; but much in the habit of walking or tripping daintily over it, on such raft or float as they may find constructed for them by water-lily or other buoyant leaves. Of these "come and trip it as you come" chicks,--(my emendation of Milton is surely more reasonable than the emendations of commentators as a body, for we do not, any of us, like to see our mistresses "trip it as they *go*")--there are, I find, pictured by Mr. Gould, three 'species,' called by him, Porzana Minuta, Olivaceous Crake; Porzana Pygmaea, Baillon's Crake; and Porzana Maruetta, Spotted Crake.

Now, in the first place, I find 'Porzana' to be indeed Italian for 'water-hen,' but I can't find its derivation; and in the second place, these little birds are neither water-hens nor moor-hens, nor water-cocks nor moor-cocks; neither can I find, either in Gould, Yarrell, or Bewick, the slightest notice of their voices!--though it is only in implied depreciation of their quality, that we have any business to call them 'Crakes,' 'Croaks,' or 'Creaks.' In the third place, 'Olivaceous' is not a translation of 'Minuta,' nor 'Baillon's' of 'Pygmaea,' nor 'spotted' of 'Maruetta'; which last is another of the words that mean nothing in any language that I know of, though the French have adopted it as 'Marouette.' And in the fourth place, I can't make out any difference, either in text or picture, between Mr. Baillon's Crake, and the 'minute' one, except that the minute one is the bigger, and has fewer white marks in the center of the back.

95. For our purposes, therefore, I mean to call all the three varieties neither Crake nor Porzan, but 'Allegretta,' which will at once remind us of their motion; the larger one, nine inches long, I find called always Spotted Crake, so that shall be 'Allegretta Maculata,' Spotty Allegret; and the two little ones shall be, one, the Tiny Allegret, and the other the Starry Allegret (Allegretta Minuta, and Allegretta Stellaris); all the three varieties being generally thought of by the plain English name I have given at the head of this section, 'Lily-Ouzel' (see, in SS 7, page 5, the explanation of my system of dual epithet, and its limitations. I note, briefly, what may be properly considered distinctive in the three kinds.)

II.A.

ALLEGRETTA NYMPHAEA, MACULATA. SPOTTED ALLEGRET.

96. Water-Crake or 'Skitty' of Bewick,--French, 'Poule d'eau Marouette,' (we may perhaps take Marouette as euphonious for Maculata, but I wish I knew what it meant);--though so light of foot, flies heavily; and, when compelled to take wing, merely passes over the tops of the reeds to some place of security a short distance off. (Gould.) The body is "in all these Rails *compressed*" (Yarrell,--he means laterally thin), which enables them to make their way through dense herbage with facility. I can't find anything clear about its country, except that it 'occasionally visits' Sweden in summer, and Smyrna in winter, and that it has been found in Corfu, Sicily, Crete,--Whittlesea Mere,--and Yarley Fen;--in marshes always, wherever it is; (nothing said of its behavior on ice,) and not generally found farther north than Cumberland. Its food is rather nasty--water-slugs and the like,--but it is itself as fat as an ortolan, "almost melts in the *hand*." (Gould.) Its own color, brown spotted with white; "the spots on the wing coverts surrounded with black, which gives them a studded or pearly appearance." (Bewick,--he means by 'pearly,' rounded or projecting.) Hence my specific epithet. Its young are of the liveliest black, "little balls of black glistening down," beautifully put by Mr. Gould among the white water Crowfoot (Ranunculus Aquatilis), looking like little ducklings in mourning. "Its nest is made of rushes and other buoyant materials matted together, so as to float on, and rise or fall with, the ebbing or flowing of the water like a boat; and to prevent its being carried away, it is moored or fastened to a reed." (Bewick.)

II.B.

ALLEGRETTA NYMPHAEA, STELLARIS. STARRY ALLEGRET.

97. Called 'Stellaris' by Temminck.--I do not find why, but it is by much the brightest in color of the three, and may be thought of as the

star of them. Gould says it is the least, also, and calls it the 'Pigmy'; but we can't keep that name without confusing it with the 'Minuta.' 'Baillon's Crake' seems the most commonly accepted title,--as the worst possible. Both this, and the more quietly toned Tiny, in Mr. Gould's delightful plates of them, have softly brown backs, exquisitely ermined by black markings at the root of each feather, following into series of small waves, like little breakers on sand. They have lovely gray chemisettes, striped gray bodices, and green bills and feet; a little orange stain at the root of the green bill, and the bright red iris of the eye have wonderful effect in warming the color of the whole bird: and with beautiful fancy Mr. Gould has put the Stellaris among yellow water-lilies to set off its gray; and a yellow butterfly with blue and red spots, and black-speckled wings (Papilio Machaon), to harmonize both. It is just as if the flower were gradually turning into the bird. Examples of the Starry Allegret *have* been 'obtained'--in the British Islands. It is said to be numerous, unobtained, in India, China, Japan, Persia, Greece, North Africa, Italy, and France. I have never heard of anybody's seeing it, however.

II.C.

ALLEGRETTA NYMPHAEA, MINUTA. TINY ALLEGRET.

98. 'Tiny Allegret,'--Yarrell's 'Little Crake,' (but see names in Appendix). It is a little more rosy than 'Stellaris' in the gray of its neck, passing into brown; and Mr. Gould has put it with a pink water plant, which harmonizes with it to the bird's advantage; while the tiny creature stands on the bent leaf of a reed, and scarcely bends it more! "It runs with rapidity over broken reeds, and moves gracefully, raising and displaying its tail at every step." It has so very small a tail to display, however, that I should hardly think the display was worth while. "It is very cunning, and especially noticeable for the subtlety with which it wearies the dog of the sportsman by executing a thousand evolutions with surprising celerity; whence comes the trivial name of 'kill-dog' bestowed upon it in some localities. Pursued to extremity, it casts itself into the water, swims with ease, and dives at the moment its enemy is about to seize it; or it conceals itself in a tuft of reeds or a bush, and by this means often escapes with impunity. It loves to breed among the reeds, and in long and thick grass, frequently in small companies of its own species, or of the Stellaris. The female lays her eggs on an inartificially constructed platform of decayed leaves or stalks of marsh plants, slightly elevated above the water." How elevated, I cannot find proper account,--that is to say, whether it is hung to the stems of

growing reeds, or built on hillocks of soil, but the bird is always liable to have its nest overflowed by floods. The full-grown bird is dressed in an exquisite perfection of barred bodice, spotted chemisette, and waved feathers edged with gray on the back.

99. The reader will please recollect these three Allegrets as the second group of the dab-or dabble-chicks; and, while the water-ouzel is a mountain and torrent bird, these inhabit exclusively flat lands and calm water, belonging properly to temperate, inclining to warm, climates, and able to gladden for us--as their name now given implies--many scenes and places otherwise little enlivened; and to make the very gnats of them profitable to us, were we wise enough. Dainty and delightful creatures in all their ways,--voice only dubitable, but I hope not a shriek or a squeak;--and there seems to be no reason whatever why half our fen lands should not be turned into beds of white water lilies and golden ducks, with jetty ducklings, to the great comfort of English souls.[22]

III.

TREPIDA STAGNARUM. LITTLE GREBE.

100. The two birds--Torrent-ouzel, and Lily-ouzel,--which we have been just describing, agree, you will observe, in delicate and singular use of their feet in the water; the torrent-ouzel holding itself mysteriously at the bottom; and the lily-ouzel, less mysteriously, but as skillfully, on the top (for I forgot to note, respecting this raft-walking, that the bird, however light, must be always careful not to tread on the edges of leaves, but in the middle, or, rather, as nearly as may be where they are set on the stalk; it would go in at once if it trod on the edges). But both the birds have the foot which is really characteristic of land, not water-birds; and especially of those land species that run well. Of the real action of the toes, either in running, or hopping, nothing is told us by the anatomists--(compare lecture on Robin, SS 26); but I hope before long to get at some of the facts respecting the greater flexibility of the gripping and climbing feet, and elasticity of running ones; and to draw up something like a properly graduated scale of the length of the toes in proportion to that of the body.

FIG. 12.

And, for one question, relative to this--the balance of a bird *standing*, not gripping--is to be thought of. Taking a typical profile of bird-form in its abstract, with beak, belly, and foot, horizontal (Fig. 12), the security of the standing, (supposing atomic weight equal through the bird's body, and the *will*, in the ankle, of iron,) is the same as of an inverted cone, between the dotted lines from the extremities of the foot to those of the body; and, of course, with a little grip of the foot or hind claw, the bird can be safe in almost any position it likes. Nevertheless, when the feet are as small in proportion as the Torrent-ouzel's, I greatly doubt the possibility of such a balance as Bewick has given it (Fig. 13 *a*). Gould's of the black-bodiced Ouzel (Fig. 13 *b*) is, I imagine, right. Bewick was infallible in plume texture, and expression either of the features of animals, or of any action that had meaning in it; but he was singularly careless of indifferent points in geometry or perspective; and even loses character in his water-birds, by making them always swim on the top of the water.

FIG. 13*a*.

101. But, whatever their balance of body, or use of foot, the two birds just examined are, as I said, essentially connected with the running land birds, or broadly, the Plovers; and with the Sand-runners, or (from their cry) Sandpipers, which Mr. Gould evidently associates mentally with the Plovers, in his description of the plumage of the Dunlin; while he gives to them in his plates of that bird--the little Stint, and common Sandpiper--most subtle action with their fine feet,--thread-fine, almost, in the toes; requiring us, it seems to me, to consider them as entirely land-birds, however fond of the wave margins. But the next real water-ouzel we come to, belongs to a group with feet like little horse-chestnut leaves; each toe having its separate lobes of web. Why separated, I cannot yet make out, but the bird swims, or even dives, on occasion, with dexterity and force. These lobe-footed birds consist first of the Grebes, which are connected with fresh-water ducks; and, secondly, of the Phalaropes, which are a sort of sea-gulls. No bird which is not properly web-footed has any business to think itself either true duck or true gull; but as, both in size and habit of life, the larger grebes and phalaropes are entirely aquatic and marine, I shall take out of them into my class of dabchicks, only those which are literally dabblers in habit, and chickens in size. And of the Grebes, therefore, only the one commonly known as the Dabchick, the 'Little Grebe,' 'Colymbus Minutus' (Minute Diver), of Linnaeus. A summary word or two, first, respecting the Grebe family, will be useful.

FIG. 13*b*.

102. Grebe, properly, I suppose, Grebe, from the French, is not in Johnson, nor do any of my books tell me what it means. I retain it, however, as being short, not ugly, and well established in two languages. We may think of it as formed from gre, and meaning 'a nice bird.' The specialities of the whole class, easily remembered, are, first, that they have chestnut-leaf feet; secondly, that their legs are serrated behind with a double row of notches--(why?); thirdly, that they have no tails; fourthly, that they have, most of them, very fine and very comic crests, tufts, tippets, and other variously applied appendages to their heads and chins, so that some are called 'crested,' some 'eared,' some 'tippeted,' and so on; but the least of them, our proper Dabchick, displays no absurdity of this sort, and I have the less scruple in distinguishing it from others. I find, further, in Stanley's classes, the Grebes placed among the short-winged birds, and made to include all the divers; but he does not say how short their wings are; and his grouping them with guillemots and puffins is entirely absurd, all their ways and looks, and abodes, being those of ducks. We can say no more of them as a family, accordingly, until we know what a duck is;--and I go on to the little pet of them, whose ways are more entirely its own.

103. Strangely, the most interesting fact (if *fact* it be) that it builds a floating nest, gains scarcely more than chance notice from its historians. Here is Mr. Gould's account of it: "The materials composing this raft or nest are weeds and aquatic plants carefully heaped together in a

rounded form; it is very large at the base, and is so constantly added to, that a considerable portion of it becomes submerged; at the same time it is sufficiently buoyant to admit of its saucer-like hollow top being always above the surface. In this wet depression five or six eggs are laid. The bird, always most alert, is still more so now, and scarcely ever admits of a near examination of the nest-making, or of a view of the eggs. In favorable situations, however, and with the aid of a telescope, the process may be watched; and it is not a little interesting to notice with what remarkable quickness the dabchick scratches the weeds over her eggs with her feet, when she perceives herself observed, so as not to lead even to the suspicion that any were deposited on the ill-shapen floating mass. This work of an instant displays as much skill in deception as can well be imagined."

104. It is still left to question, first, what is meant by a wet depression?--does the bird actually sit in the water, and are the eggs under it? and, if not, how is the water kept out? Secondly, is the floating nest anchored, and how? Looking to other ornithologists for solution of these particulars, I find nobody else say anything about a floating nest at all. Bewick describes it as being of a large size, and composed of a very great quantity of grass and water plants, at least a foot in thickness, and so placed in the water that the female hatches her eggs amidst the continual wet in which they were first laid. Yarrell says only that it is a large flat nest made of aquatic plants; while Morris finally complicates the whole business by telling us that the nest is placed often as much as twenty or thirty yards from the water, that it is composed of short pieces of roots, reeds, rushes, and flags, and that when dry the whole naturally becomes very brittle.[23]

105. While, out of my fifteen volumes of ornithology, I can obtain only this very vague account of the prettiest bird, next to the kingfisher, that haunts our English rivers, I have no doubt the most precise and accurate accounts are obtainable of the shapes of her bones and the sinuosities of her larynx; but about these I am low-minded enough not to feel the slightest curiosity. I return to Mr. Gould, therefore, to gather some pleasanter particulars; first, namely, that she has a winter and summer dress,--in winter olive gray and white, but in summer, (changing at marriage time) deep olive black, with dark chestnut chemisette. Infant dabchicks have "delicate rose-colored bills, harlequin-like markings, and rosy-white aprons." The harlequin-like markings I should call, rather, agate-like, especially on the head, where they are black and white, like

an onyx. The bodies look more like a little walnut-shell, or nutmeg with wings to it, or things that are to be wings, some day.

106. Even when full-grown, the birds never fly much,--never more, says Morris, "than six or ten feet above the water, and for the most part trailing their legs in it; but either on the water or under it, every movement is characterized by the most consummate dexterity, and facile agility. The most expert waterman that sculls his skiff on the Thames or Isis, is but an humble and unskillful imitator of the dabchick. In moving straightforward (under water?), the wings are used to aid its progress, as if in the air, and in turning it has an easy gliding motion, feet and wings being used, as occasion requires, sometimes on one side and sometimes on the other. It walks but indifferently, as may readily be imagined from the position of the legs, so very far back. It is pleasant to watch the parent bird feeding her young: down she dives with a quick turn, and presently rises again with, five times out of six, a minnow, or other little fish, glittering like silver in her bill. The young rush towards the spot where the mother has come up, but she does not drop the fish into the water for them to receive until she has well shaken it about and killed it, so that it may not escape, when for the last time in its own element. I have seen a young one which had just seized, out of its turn I have no doubt, the captured prey, chased away by her, and pursued in apparent anger, as if for punishment, the following one being willingly given the next fish without any demur."

107. Mr. Gould seems to think that the dabchick likes insects and fish spawn better than fish, or at least more prudently dines upon them. "That fish are taken we have positive evidence from examples having been repeatedly picked up dead by the fishermen of the Thames, with a bull-head or miller's thumb in their throats, and by which they had evidently been choked in the act of swallowing them. That it is especially fond of insects is shown by the great activity it displays, when in captivity, in capturing house-flies and other diptera. Those who have visited Paris will probably have seen the grebes in the window of the restaurateur in the Rue de Rivoli. For years have a pair of these birds been living, apparently in the greatest enjoyment, within the glass window, attracting the admiration of all the passers-by. The extreme agility with which they sailed round their little prison, or scrambled over the half-submerged piece of rock for a fly, was very remarkable. That no bird can be more easily kept in a state of confinement is certain."

108. This question about its food is closely connected with that of its diving. So far as I understand Mr. Morris, it dives only when disturbed,

and to escape,--remaining under water, however, if need be, an almost incredible time, and swimming underneath it to great distances. Here we have, if we would only think of it, the same question as that about the water-ouzel, how it *keeps down*; and we must now note a few general points about diving birds altogether.

It is easy to understand how the properly so-called divers can plunge with impetus to great depths, or keep themselves at the bottom by continued strokes of the webbed feet; but neither how the ouzel walks at the bottom, if it be specifically lighter than the water, nor how a bird can swim horizontally under the surface; at least it is not enough explained that the action must be always that of oblique diving, the bird regulating the stroke according to the upward pressure of the water at different depths.

109. But there are many other points needing elucidation. It is said (and beautifully insisted on, by Michelet,) that great spaces in the bones of birds that pass most of their lives in flight are filled with air: presumably the bones of the divers are made comparatively solid, or it is even conceivable--if conceptions or suppositions were of any use,--that the deep divers may take in water, to help themselves to sink. The enormous depths at which they have been caught, according to report, cannot be reached by any mere effort of strength, if the body remained as buoyant as it evidently is on the surface. The strength of the wing must, however, be enormous, for the great northern diver is described as swimming under water "as it were with the velocity of an arrow in the air" (Yarrell, vol. iii., page 431); or to keep to more measured fact, Sir William Jardine says, "I have pursued this bird in a Newhaven fishing-boat with four sturdy rowers, and notwithstanding it was kept almost constantly under water by firing as soon as it appeared, the boat could not succeed in making one yard upon it" (*ibid.*, p. 432).

110. But this is followed by the amazing statement of Mr. Robert Dunn, p. 433, that in the act of diving it does not appear to make the least exertion, but sinks gradually under the surface, without throwing itself forward, the head being the last part that disappears. I am not fond of the word 'impossible,' but I think I am safe in saying that according to the laws of nature no buoyant body can sink merely by an act of volition; and that it must pull itself down by some hitherto unconceived action of the feet, which in this bird are immensely broad and strong, and so flat that it cannot walk with them, any more than we could with two flat boards a yard square tied to our feet; but, when it is caught on land, shoves its body along upon the ground, like a seal, by jerks. All these

diving motions are executed in a more delicate but quite as wonderful way by the dabchick,--more wonderful indeed it may be said, because it has only the divided or chestnut-leaf-like foot, to strike with. We shall understand it perhaps a little better after tracing, in a future talk, the history of its relations among the smaller sea-gulls; meantime, in quitting the little dainty creature, I must plead for a daintier Latin name than it has now--'Podiceps.' No one seems to have the least idea what that means; and 'Colymbus,' diver, must be kept for the great Northern Diver and his deep-sea relatives, far removed from our little living ripple-line of the pools. I can't think of any one pretty enough; but for the present 'Trepida' may serve; and perhaps be applied, not improperly, to all the Grebes, with reference to their subtle and instant escape from any sudden danger. (See Stanley, p. 419.) "It requires all the address of a keen sportsman to get within shot," and when he does, the bird may still be too shrewd for him. "I fired at the distance of thirty yards; my gun went quick as lightning, but the grebe went quicker, and scrambling over, out of sight, came up again in a few seconds perfectly unhurt."

I think, therefore, that unless I receive some better suggestion, 'Trepida Stagnarum' may be the sufficiently intelligible Latin renaming of our easily startled favorite.

IV.

TITANIA ARCTICA. ARCTIC FAIRY.

111. I must first get quit of the confusion of names for this bird. Linnaeus, in the Fauna Suecica, p. 64, calls it 'Tringa Lobata,' but afterwards 'Northern Tringa'; and his editor, Gmelin, 'Dark Tringa.' Other people agree to call it a 'phalarope,' but some of them 'northern' phalarope, some, the 'dark' phalarope; some, the 'ashy' phalarope, some, the 'disposed to be ashy' phalarope; some, the 'red-necked' phalarope; and some, 'Mr. Williams's' phalarope; finally, Cuvier calls it a 'Lobipes,' and Mr. Gould, in English, 'red-necked phalarope.' Few people are likely to know what 'Phalarope' means,[24] and I believe nobody knows what 'Tringa' means; and as, also, nobody ever sees it, the little bird being obliged to live in Orkney, Greenland, Norway, and Lapland, out of human creatures' way, I shall myself call it the Arctic Fairy. It would come south if we would let it, but of course Mr. Bond says, "The first specimen I ever had was shot by a friend of mine in September, 1842, near Southend, Essex, where he saw the phalarope swimming on the water, like a little duck, about a mile from land; not knowing what it was, he shot it, and kindly brought it to me." Another was shot while

running between the metals of the Great Eastern Railway, near the Stratford station, early in June, 1852; and on the Norfolk coast, four others have been killed during the last fifteen years; and the birds' visits, thus, satisfactorily, put a stop to. I can therefore study it only in Mr. Gould's drawing, on consulting which, I find the bird to be simply a sea dabchick,--brown stripes on the back, and all; but the webs of the feet a little finer, and in its habits it is more like the Lily-ouzel, according to the following report of Mr. St. John: "The red-necked phalarope is certainly the most beautiful little wader of my acquaintance. There were a pair of them, male and female, feeding near the loch, in a little pool which was covered with weeds of different kinds. Nothing could be more graceful than the movements of these two little birds, as they swam about in search of insects, *etc.* Sometimes *they ran lightly on the broad leaves of the water-lily which served them for a raft*, and entirely kept them out of the water. Though not exactly web-footed, the phalarope swims with the greatest ease. The attachment of these two birds to each other seemed very great: whenever in their search for food they wandered so far apart as to be hidden by the intervening weeds, the male bird stopped feeding suddenly, and, looking round, uttered a low and musical call of inquiry, which was immediately answered by the female in a different note, but perfectly expressive of her answer, which one might suppose to be to the purport that she was at hand and quite safe; on hearing her, the male immediately recommenced feeding, but at the same time making his way towards her; she also flew to meet him; they then joined company for a moment or two, and, after a few little notes of endearment, turned off again in different directions. This scene was repeated a dozen times while I was watching them. They seemed to have not the slightest fear of me, for frequently they came to within a yard of where I was sitting, and after looking up they continued catching the small water-insects, etc., on the weeds, without minding my presence in the least." What reward the birds got for this gentle behavior, we learn from the sentence following after the next two lines, containing the extremely valuable contribution to their natural history, that "on dissecting the female we found two eggs in her."

112. All other accounts concur in expressing (with as much admiration as is possible to naturalists) the kindly and frank disposition of this bird; which for the rest is almost a central type of all bird power with elf gifts added: it flies like a lark, trips on water-lily leaves like a fairy, swims like a duck, and roves like a sea-gull, having been seen sixty miles from land: and, finally, though living chiefly in Lapland and Iceland, and other such

northern countries, it has been seen serenely swimming and catching flies in the hot water of the geysers, in which a man could not bear his hand.

And no less harmoniously than in report of the extreme tameness, grace, and affectionateness of this bird do sportsmen agree also in the treatment and appreciation of these qualities. Thus says Mr. Salmon: "Although we shot two pairs, those that were swimming about did not take the least notice of the report of the gun, and they seemed to be much attached to each other; for when one of them flew to a short distance, the other directly followed; and while I held a wounded female in my hand, its mate came and fluttered before my face." (Compare the scene between Irene and Hector, at page 393 of the May number of *Aunt Judy's Magazine*.) And, again, says Mr. Wolley: "The bird is extremely tame, swimming about my india-rubber boat so near that I could almost catch it in my hand; I have seen it even, when far from its nest, struck at many times with an oar before it flew away." In its domestic habits also the creature seems as exemplary as, in its social habits, it is frank; for on the approach of danger to her nestlings, the hen uses all the careful subtleties of the most cunning land birds, "spreading her wings, and counterfeiting lameness, for the purpose of deluding the intruder; and after leading the enemy from her young, she takes wing and flies to a great height, at the same time displaying a peculiar action of the wings; then descending with great velocity, and making simultaneously a noise with her wings. On her return to her young, she uses a particular cry for the purpose of gathering them together. As soon as she has collected them, she covers them with her wings, like the domestic hen."

113. I cannot quite make out the limits of the fairy's migrations; but it is said by Morris to 'occur' in France, Holland, Germany, Italy, and Switzerland. I find that one was what sportsmen call 'procured' near York, in full summer dress; and another killed at Rottingdean, swimming in a pond in the middle of the village, in the company of some ducks. At Scarborough, Louth, and Shoreham, it has also been captured or shot, and has been 'found' building nests in Sutherland: and, on the whole, it seems that here is a sort of petrel-partridge, and duckling-dove, and diving-lark, with every possible grace and faculty that bird can have, in body and soul; ready, at least in summer, to swim on our village ponds, or, wait at our railway stations, and make the wild north-eastern coasts of Scotland gay with its dancing flocks upon the foam; were it not that the idle cockneys, and pot-headed squires fresh out of Parliament, stand as it were on guard all round the island,

spluttering small-shot at it, striking at it with oars, cutting it open to find how many eggs there are inside, and, in fine, sending it for refuge into the hot water of Hecla, and any manner of stormy solitude that it can still find for itself and its amber nestlings. I have never seen one, nor I suppose ever shall see, but hear of some of my friends sunning themselves at midnight about the North Cape, of whom, if any one will bring me a couple of Arctic fairies in a basket, I think I can pledge our own Squire's and Squire's lady's faith, for the pair's getting some peace, if they choose to take it, and as many water-lily leaves as they can trip upon, on the tarns of Monk-Coniston.

IV.B.

TITANIA INCONSTANS. CHANGEFUL FAIRY.

Phalaropus Fulicarius. (*Coot-like Phalarope--Gould.*)

114. I think the epithet 'changeful' prettier, and, until we know what a coot *is* like, more descriptive, than 'coot-like'; the bird having red plumage in summer, and gray in winter, while the coot is always black. It is a little less pretty and less amiable than its sister fairy; otherwise scarcely to be thought of but as a variety, both of them being distinguished from the coot, not only by color, but by their smaller size;--(they eight inches long, it sixteen)--and by the slender beaks, the coot having a thick one, half-way to a puffin's.

And here, once for all,--for I see I have taken no note yet of the beaks or bills of my dabchicks,--I will at once arrange a formula of the order of questions which it will be proper to ask, and get answered, concerning any bird, in the same order always, so that we shall never miss anything that we ought to think of. And I find these questions will naturally and easily fall into the following twelve:

1. Country, and scope of migration.
2. Food.
3. Form and flight.
4. Foot.
5. Beak and eye.
6. Voice and ear.
7. Temper.
8. Nest.
9. Eggs.
10. Brood.
11. Feathers.
12. Uses in the world.

It may be thought that I have forced--and not fallen into--my number 12, by packing the faculties of sight and hearing into by-corners. But the expression of a bird's head depends on the relation of eye to beak, as the getting of its food depends on their practical alliance of power; and the question, for instance, whether peacocks and parrots have musical ears, seems to me not properly debatable unless with due respect to the quality of their voices. It is curious, considering how much, one way or another, we are amused or pleased by the chatter and song of birds, that you will scarcely find in any ornithic manual more than a sentence, if so much, about their hearing; and I have not myself, at this moment, the least idea where a nightingale's ears are! But see Appendix, p. 122.

I retain, therefore, my dodecahedric form of catechism as sufficiently clear; and without binding myself to follow the order of it in strictness, if there be motive for discursory remark, it will certainly prevent my leaving any bird insufficiently distinguished, and enable me to arrange the collected statements about it in the most easily compared order.

115. We will try it at once on this second variety of the Titania, of which I find nothing of much interest in my books, and have nothing discursive myself to say.

1. Country. Arctic mostly; seen off Greenland, in lat. 68deg, swimming among icebergs three or four miles from shore. Abundant in Siberia, and as far south as the Caspian. Migratory in Europe as far as Italy, yet always rare. (Do a few only, more intelligently curious than the rest, or for the sake of their health, travel?)

2. Food. Small thin-skinned crustacea, and aquatic surface-insects.

3. Form and flight. Stout, for a sea-bird; and they don't care to fly, preferring to *swim* out of danger. Body 7 to 8 inches long; wings, from carpal joint to end, 43/4,--say 5. These quarters of inches, are absurd pretenses to generalize what varies in every bird. 8 inches long, by 10 across the wings open, is near enough. In future, the brief notification 8 x 10, 5 x 7, or the like, will enough express a bird's inches, unless it possess decorative appendage of tail, which must be noted separately.

4. Foot. Chestnut-leaved in front toes, the lobes slightly serrated on the edges. Hind toe without membrane. Color of foot, always black.

5. Beak. Long, slender, straight. (How long? Drawn as about a fifth of the bird's length--say an inch, or a little over.) Upper mandible slightly curved down at the point. In Titania arctica, the beak is longer and more slender.

6. Voice. A sharp, short cry, not conceived by me enough to spell any likeness of it.

7. Temper. Gentle, passing into stupid, (it seems to me); one, in meditative travel, lets itself be knocked down by a gardener with his spade.

8. Nest. Little said of it, the bird breeding chiefly in the North. Among marshes, it is of weeds and grass; but among icebergs, of what?

9. Eggs. Pear-shape; narrow ends together in nest; never more than four.

10. Brood. No account of.

11. Feathers. Mostly gray, passing into brown in summer, varied with white on margin. Reddish chestnut or bay bodice--well oiled or varnished.

12. Uses. Fortunately, at present, unknown.

V.

RALLUS AQUATICUS. WATER-RAIL.

116. Thus far, we have got for representatives of our dabchick group, eight species of little birds--namely, two Torrent-ouzels, three Lily-ouzels, one Grebe, and two Titanias. And these we associate, observe, not for any specialty of feature in them, but for common character, habit, and size; so that, if perchance a child playing by any stream, or on the sea-sands, perceives a companionable bird dabbling in an equally childish and pleasant manner, he may not have to look through half a dozen volumes of ornithology to find it; but may be pretty sure it has been one of these eight. And having once fastened the characters of these well in his mind, he may with ease remember that the little grebe is the least of a family of chestnut-leaf-footed, and sharp-billed creatures, which yet in size, color, and diving power, go necessarily among Ducks, and cannot be classed with Dabblers; though it must be always as distinctly kept in mind that a duck *proper* has a flat beak, and a fully webbed foot.

Again, he may recollect that with these leaf-footed ducks of the calm and fresh waters, must be associated the leaf-footed or fringefooted ducks of the sea;--'phalaropes,' which by their short wings connect themselves with many clumsy marine creatures, on their way to become seals instead of birds; and that I have kept the two little Titanias out of this class, not merely for their niceness, but because they are not short-winged in any vulgar degree, but seem to have wings about as long as a sandpiper's;--and indeed I had put the purple sandpiper, Arquatella maritima, with them, in my own folio; only as the Arquatella's feet are not chestnutty, she had better go with her own kind in our notes on them.

117. But there are yet two birds, which I think well to put with our eight dabchicks, though they are much larger than any of them,--partly because of their disposition, and partly because of their plumage,--the water-rail, and water-hen. Modern science, with instinctive horror of all that is pretty to see, or easy to remember, entirely rejects the plumage, as any element or noticeable condition of bird-kinds; nor have I ever yet tried to make it one myself; yet there are certain qualities of downiness in ducks, fluffiness in owls, spottiness in thrushes, patchiness in pies, bronzed or rusty luster in cocks, and pearly iridescence in doves, which I believe may be aptly brought into connection with other defining characters; and when we find an entirely similar disposition of plumage, and nearly the same form, in two birds, I do not think that *mere* difference in size should far separate them.

Bewick, accordingly, calls the water-rail the 'Brook-ouzel,' and puts it between the little crake and the water-ouzel; but he does not say a word of its living by brooks,--only 'in low wet places.' Buffon, however, takes it with the land-rail; Gould and Yarrell put it between the little crake and water-hen. Gould's description of it is by no means clear to me:--he first says it is, in action, as much "like a rat as a bird;" then that it "bounds like a ball," (before the nose of the spaniel); and lastly, in the next sentence, speaks of it as "this *lath*-like bird"! It is as large as a bantam, but can run, like the Allegretta, on floating leaves; itself, weighing about four ounces and a half (Bewick), and rarely uses the wing, flying very slowly. I imagine the 'lath-like' must mean, like the more frequent epithet 'compressed,' that the bird's body is vertically thin, so as to go easily between close reeds.

118. We will try our twelve questions again.

1. Country. Equally numerous in every part of Europe, in Africa, India, China, and Japan; yet hardly anybody seems to have seen it. Living, however, "near the perennial fountains" (wherever those may be;--it sounds like the garden of Eden!) "during the greater part of the winter, the birds pass Malta in spring and autumn, and have been seen fifty leagues at sea off the coast of Portugal" (Buffon); but where coming from, or going to, is not told. Tunis is the most southerly place named by Yarrell.

2. Food. Anything small enough to be swallowed, that lives in mud or water.

3. Form and flight. I am puzzled, as aforesaid, between its likeness to a ball, and a lath. Flies heavily and unwillingly, hanging its legs down.

4. Foot. Long-toed and flexible.

5. Beak. Sharp and strong, some inch and a half long, showing distinctly the cimeter-curve of a gull's, near the point.

6. Voice. No account of.

7. Temper. Quite easily tamable, though naturally shy. Feeds out of the hand in a day or two, if fed regularly in confinement.

8. Nest. "Slight, of leaves and strips of flags" (Gould); "of sedge and grass, rarely found," (Yarrell). Size not told.

9. Eggs. Eight or nine! cream-white, with rosy yolk!! rather larger than a blackbird's!!!

10. Brood. Velvet black, with white bills; hunting with the utmost activity from the minute they are hatched.

11. Feathers. Brown on the back, a beautiful warm ash gray on the breast, and under the wings transverse stripes of very dark gray and white. The disposition of pattern is almost exactly the same as in the Allegretta.

12. Uses. By many thought delicious eating. (Bewick.) The fact is, or seems to me, that this entire group of marsh birds is meant to become to us the domestic poultry of marshy land; and I imagine that by proper irrigation and care, many districts of otherwise useless bog and sand, might be made more profitable to us than many fishing-grounds.

VI.

PULLA AQUATICA. WATER-HEN.

(*Gallinula Chloropus.--Pennant, Bewick, Gould, and Yarrell.*)

119. 'Green-footed little cock, or hen,' that is to say, in English; only observe, if you call the Fringefoot a Phalarope, you ought in consistency to call the Green-foot a Chlorope. Their feet are not only notable for greenness, but for size: they are very ugly, having the awkward and illused look of the feet of Scratchers, while a trace of beginning membrane connects them with the fringefoots.

Their proper name would be Marsh-cock, which would enough distinguish them from the true Moor-cock or Black-cock. 'Moat-cock' would be prettier, and characteristic; for in the old English days they used to live much in the moats of manor-houses; mine is the name nearest to the familiar one; only note there is no proper feminine of 'pullus,' and I use the adjective 'pulla' to express the dark color.

It is a dark-*brown* bird, according to the colored pictures--iron *gray*, Buffon says, with white stripes of little order on the bodice, clumsy feet and bill, but makes up for all ungainliness by its gentle and intelligent mind; and seems meant for a useful possession to mankind all over the world, for it lives in Siberia and New Zealand; in Senegal and Jamaica; in Scotland, Switzerland, and Prussia; in Corfu, Crete, and Trebizond; in Canada, and at the Cape. I find no account of its migrations, and one would think that a bird which usually flies "dip, dip, dipping with its toes, and leaving a track along the water like that of a stone at 'ducks and drakes'" (Yarrell), would not willingly adventure itself on the Atlantic. It must have a kind of human facility in adapting itself to climate, as it has human domesticity of temper, with curious fineness of sagacity and sympathies in taste. A family of them, petted by a clergyman's wife, were constantly adding materials to their nest, and "made real havoc in the flower-garden,--for though straw and leaves are their chief ingredients, they seem to have an eye for beauty, and the old hen has been seen surrounded with a brilliant wreath of scarlet anemones." Thus Bishop Stanley, whose account of the bird is full of interesting particulars. This aesthetic water-hen, with her husband, lived at Cheadle, in Staffordshire, in the rectory moat, for several seasons, "always however leaving it in the spring," (for Scotland, supposably?): being constantly fed, the pair became quite tame, built their nest in a thorn-bush covered with ivy which had fallen into the water; and "when the young are a few days old, the old ones bring them up close to the drawing-room window, where they are regularly fed with wheat; and, as the lady of the house pays them the greatest attention, they have learned to look up to her as their natural protectress and friend; so much so, that one bird in particular, which was much persecuted by the rest, would, when attacked, fly to her for refuge; and whenever she calls, the whole flock, as tame as barn-door fowls, quit the water, and assemble round her, to the number of seventeen. (November, 1833.)

120. "They have also made other friends in the dogs belonging to the family, approaching them without fear, though hurrying off with great alarm on the appearance of a strange dog.

"The position of the water, together with the familiarity of these birds, has afforded many interesting particulars respecting their habits.

"They have three broods in a season--the first early in April; and they begin to lay again when the first hatch is about a fortnight old. They lay eight or nine eggs, and sit about three weeks,--the cock alternately with the hen. The nest in the thorn-bush is placed usually so high above the surface of the water, they cannot climb into it again; but, as a substitute, within an hour after they leave the nest, the cock bird builds a larger and more roomy nest for them, with sedges, at the water's edge, which they can enter or retire from at pleasure. For about a month they are fed by the old birds, but soon become very active in taking flies and water-insects. Immediately on the second hatch coming out, the young ones of the first hatch assist the old ones in feeding and hovering over them, leading them out in detached parties, and making additional nests for them, similar to their own, on the brink of the moat.

"But it is not only in their instinctive attachments and habits that they merit notice; the following anecdote proves that they are gifted with a sense of observation approaching to something very like reasoning faculties.

"At a gentleman's house in Staffordshire, the pheasants are fed out of one of those boxes described in page 287, the lid of which rises with the pressure of the pheasant standing on the rail in front of the box. A water-hen observing this, went and stood upon the rail as soon as the pheasant had quitted it; but the weight of the bird being insufficient to raise the lid of the box, so as to enable it to get at the corn, the water-hen kept jumping on the rail to give additional impetus to its weight: this partially succeeded, but not to the satisfaction of the sagacious bird. Accordingly it went off, and soon returning with a bird of its own species, the united weight of the two had the desired effect, and the successful pair enjoyed the benefit of their ingenuity.

"We can vouch for the truth of this singular instance of penetration, on the authority of the owner of the place where it occurred, and who witnessed the fact."

121. But although in these sagacities, and teachablenesses, the bird has much in common with land poultry, it seems not a link between these and water-fowl; but to be properly placed by the ornithologists between the rail and the coot: this latter being the largest of the fringefoots, singularly dark in color, and called 'fulica' (sooty), or, with insistence, 'fulica atra' (black sooty), or even 'fulica aterrima' (blackest sooty). 'Coot' is said by Johnson to be Dutch; and that it became 'cotee' in French; but I cannot find cotee in my French dictionary. In the meantime, putting the coot and water-hen aside for future better knowledge, we may be content with the pentagonal group of our dabchicks--passing at each angle into another tribe, thus,--(if people must classify, they at least should also *map*). Take the Ouzel, Allegret, Grebe, Fairy, and Rail, and, only giving the Fairy her Latin name, write their fourpenny-worth of

initial letters (groat) round a pentagon set on its base, putting the Ouzel at the top angle,--so. Then, the Ouzels pass up into Blackbirds, the Rails to the left into Woodcocks, the Allegrets to the right into Plovers, the Grebes, down left, into Ducks, and the Titanias, down right, into Gulls. And *there's* a bit of pentagonal Darwinism for you, if you like it, and learn it, which will be really good for something in the end, or the five ends.

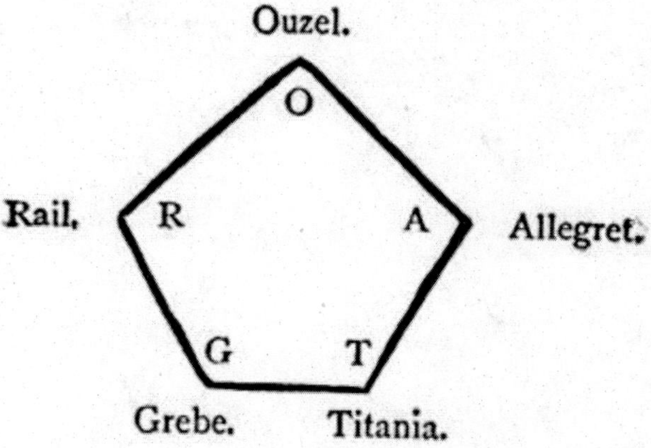

122. And for the bliss of classification pure, with no ends of any sort or any number, referring my reader to the works of ornithologists in general, and for what small portion of them he may afterwards care to consult, to my Appendix, I will end this lecture, and this volume, with the refreshment for us of a piece of perfect English and exquisite wit, falling into verse,--the Chorus of the Birds, in Mr. Courthope's Paradise of them,--a book lovely, and often faultless, in most of its execution, but little skilled or attractive in plan, and too thoughtful to be understood without such notes as a good author will not write on his own work; partly because he has not time, and partly because he always feels that if people won't look for his meaning, they should not be told it. My own special function, on the contrary, is, and always has been, that of the Interpreter only, in the 'Pilgrim's Progress;' and I trust that Mr. Courthope will therefore forgive my arranging his long cadence of continuous line so as to come symmetrically into my own page, (thus also enforcing, for the inattentive, the rhymes which he is too easily proud to insist on,) and my division of the whole chorus into equal strophe and antistrophe of six lines each, in which, counting from the

last line of the stanza, the reader can easily catch the word to which my note refers.

123.

We wish to declare,
            How the birds of the air
All high institutions designed,
And, holding in awe
            Art, Science, and Law,
Delivered the same to mankind.
To begin with; of old
            Man went naked, and cold,
Whenever it pelted or froze,
Till *we* showed him how feathers
            Were proof against weathers,
With that, *he* bethought him of hose.
And next, it was plain,
            That he, in the rain,
Was forced to sit dripping and blind,
While the Reed-warbler swung
            In a nest, with her young
Deep sheltered, and warm, from the wind.
So our homes in the boughs
            Made *him* think of the House;
And the Swallow, to help him invent,Revealed the best way
            To economize clay,
And bricks to combine with cement.
The knowledge withal
            Of the Carpenter's awl,
Is drawn from the Nuthatch's bill;
And the Sand-Martin's pains
            In the hazel-clad lanes
Instructed the Mason to drill.
Is there *one* of the Arts,
            More dear to men's hearts?
To the bird's inspiration they owe it;
For the Nightingale first
            Sweet music rehearsed,
Prima-Donna, Composer, and Poet.
The Owl's dark retreats

Showed sages the sweets
Of brooding, to spin, or unravel
Fine webs in one's brain,
Philosophical--vain;
The Swallows,--the pleasures of travel.
Who chirped in such strain
Of Greece, Italy, Spain
And Egypt, that men, when they heard,
Were mad to fly forth,
From their nests in the North,
And follow--the tail of the Bird.
Besides, it is true,
To *our* wisdom is due
The knowledge of Sciences all;
And chiefly, those rareMetaphysics of Air
Men 'Meteorology' call,
And men, in their words,
Acknowledge the Birds'
Erudition in weather and star;
For they say, "'Twill be dry,--
The swallow is high,"
Or, "Rain, for the Chough is afar."
'Twas the Rooks who taught men
Vast pamphlets to pen
Upon social compact and law,
And Parliaments hold,
As themselves did of old,
Exclaiming 'Hear, Hear,' for 'Caw, Caw.'
And whence arose Love?
Go, ask of the Dove,
Or behold how the Titmouse, unresting,
Still early and late
Ever sings by his mate,
To lighten her labors of nesting.
Their bonds never gall,
Though the leaves shoot, and fall,
And the seasons roll round in their course,
For their marriage, each year,
Grows more lovely and dear;
And they know not decrees of Divorce.

That these things are truth
We have learned from our youth,
For our hearts to our customs incline,
As the rivers that roll
From the fount of our soul,
Immortal, unchanging, divine.
Man, simple and old,
In his ages of gold,
Derived from our teaching true light,
And deemed it his praise
In his ancestors' ways
To govern his footsteps aright.
But the fountain of woes,
Philosophy, rose;
And, what between reason and whim,
He has splintered our rules
Into sections and schools,
So the world is made bitter, for *him*.
But the birds, since on earth
They discovered the worth
Of their souls, and resolved with a vow
No custom to change,
For a new, or a strange,
Have attained unto Paradise, *now*.

Line 9. PELTED, said of *hail*, not rain. Felt by nakedness, in a more severe manner than mere rain.

11. 'WEATHERS,' *i.e., both* weathers--hail and cold: the *armor* of the feathers against hail; the down of them against cold. See account of Feather-mail in 'Laws of Fesole,' chap, vi., p. 53, with the first and fifth plates, and figure 15.

15. BLIND. By the beating of the rain in his face. In *hail*, there is real danger and bruising, if the hail be worth calling so, for the whole body; while in rain, if *it* be rain also worth calling rain, the great plague is the beating and drenching in the face.

16. SWUNG. Opposed to 'sit' in previous line. The human creature, though it sate steady on this unshakable earth, had no house over its head. The bird, that lived on the tremblingest and weakest of bending things, had her *nest* on it, in which even her infinitely tender brood were *deep* sheltered and warm, from the *wind*. It is impossible to find a lovelier instance of pure poetical antithesis.

20. HOUSE. Again antithetic to the perfect word 'Home' in the line before. A house is exactly, and only, half-way to a 'home.' Man had not yet got so far as even that! and had lost, the chorus satirically imply, even the power of getting the other half, ever, since his "*She* gave me of the tree."

24. BRICKS. The first bad inversion permitted, for "to combine bricks with cement." In my Swallow lecture I had no time to go into the question of her building materials; the point is, however, touched upon in the Appendix (pp. 110, 112, and note).

30. 'DRILL,' for 'quarry out,' 'tunnel,' etc., the best general term available.

36. COMPOSER of the music; POET of the meaning.

Compare, and think over, the Bullfinch's nest, etc., SS 48 to 61 of 'Eagle's Nest.'

In modern music the *meaning* is, I believe, by the reputed masters omitted.

39. To SPIN, or *un*ravel. Synthesis and analysis, in the vulgar Greek slang.

46. MAD. Compare Byron of the English in *his* day. "A parcel of staring boobies who go about gaping and wishing to be at once cheap and magnificent. A man is a fool now, who travels in France or Italy, till that tribe of wretches be swept home again. In two or three years, the first rush will be over, and the Continent will be roomy and agreeable." (Life, vol. ii., p. 319.) For sketches of the English of seventeen years later, at the same *spots* (Wengern Alp and Interlachen), see, if you *can* see, in any library, public or private, at Geneva, Topffer's 'Excursions dans les Alpes, 1832.' Douzieme, Treizieme, and Quatorzieme Journee.

48. THE TAIL. Mr. Courthope does not condescend to italicize his pun; but a swallow-tailed and adder-tongued pun like this must be paused upon. Compare Mr. Murray's Tale of the Town of Lucca, to be seen between the arrival of one train and the departure of the next,--nothing there but twelve churches and a cathedral,--mostly of the tenth to thirteenth century.

60. AFAR. I did not know of this weather sign; nor, I suppose, did the Duke of Hamilton's keeper, who shot the last pair of Choughs on Arran in 1863. ('Birds of the West of Scotland,' p. 165.) I trust the climate has wept for them; certainly our Coniston clouds grow heavier, in these last years.

63. SOCIAL. Rightly sung by the Birds in three syllables; but the lagging of the previous line (probably intentional, but not pleasant,)

makes the lightness of this one a little dangerous for a clumsy reader. The 'i-al' of 'social' does not fill the line as two full short syllables, else the preceding word should have been written '*on,*' not 'upon.' The five syllables, rightly given, just take the time of two iambs; but there *are* readers rude enough to accent the 'on' of upon, and take 'social' for two short syllables.

64. HOLD. Short for 'to hold'--but it is a licentious construction, so also, in next line, 'themselves' for 'they themselves.' The stanza is on the whole the worst in the poem, its irony and essential force being much dimmed by obscure expression, and even slightly staggering continuity of thought. The Rooks may be properly supposed to have taught men to dispute, but not to write. The Swallow teaches building, literally, and the Owl moping, literally; but the Rook does not teach pamphleteering literally. And the 'of old' is redundant, for rhyme's sake, since Rooks hold parliaments now as much as ever they did.

76. EACH YEAR. I doubt the fact; and too sadly suspect that birds take different mates. What a question to have to ask at this time of day and year!

82. RIVERS. Read slowly. The 'customs' are rivers that 'go on forever' flowing from the fount of the soul. The Heart drinks of them, as of waterbrooks.

92. PHILOSOPHY. The author should at least have given a note or two to explain the sense in which he uses words so wide as this. The philosophy which begins in pride, and concludes in malice, is indeed *a* fountain--though not *the* fountain--of woes, to mankind. But true philosophy such as Fenelon's or Sir Thomas More's, is a well of peace.

98. WORTH. Again, it is not clearly told us what the author means by the worth of a bird's soul, nor how the birds learned it. The reader is left to discern, and collect for himself--with patience such as not one in a thousand now-a-days possesses, the opposition between the "fount of our soul" (line 83) and fountain of philosophy.

124. I could willingly enlarge on these last two stanzas, but think my duty will be better done to the poet if I quote, for conclusion, two lighter pieces of his verse, which will require no comment, and are closer to our present purpose. The first,--the lament of the French Cook in purgatory,--has, for once, a note by the author, giving M. Soyer's authority for the items of the great dish,--"symbol of philanthropy, served at York during the great commemorative banquet after the first exhibition." The commemorative soul of the tormented Chef--always making a dish like it, of which nobody ever eats--sings thus:--

"Do you veesh
To hear before you taste, of de hundred-guinea deesh?
Has it not been sung by every knife and fork,
'L'extravagance culinaire a l'Alderman,' at York?
Vy, ven I came here, eighteen Octobers seence,
I dis deesh was making for your Royal Preence,
Ven half de leeving world, cooking all de others,
Swore an oath hereafter, to be men and brothers.
All de leetle Songsters in de voods dat build,
Hopped into the kitchen asking to be kill'd;
All who in de open furrows find de seeds,
Or de mountain berries, all de farmyard breeds,--
Ha--I see de knife, vile de deesh it shapens,
Vith les petits noix, of four-and-twenty capons,
Dere vere dindons, fatted poulets, fowls in plenty,
Five times nine of partridges, and of pheasants twenty;
Ten grouse, that should have had as many covers,
All in dis one deesh, with six preety plovers,
Forty woodcocks, plump, and heavy in the scales,
Pigeons dree good dozens, six-and-dirty quails,
Ortulans, ma foi, and a century of snipes,
But de preetiest of dem all was twice tree dozen pipes
Of de melodious larks, vich each did clap the ving,
And veeshed de pie vas open, dat dey all might sing!"

125. There are stiff bits of prosody in these verses,--one or two, indeed, quite unmanageable,--but we must remember that French meter will not read into ours. The last piece I will give flows very differently. It is in express imitation of Scott--but no nobler model could be chosen; and how much better for minor poets sometimes to write in another's manner, than always to imitate their own.

This chant is sung by the soul of the Francesca of the Bird-ordained purgatory; whose torment is to be dressed only in falling snow, each flake striking cold to her heart as it falls,--but such lace investiture costing, not a cruel price per yard in souls of women, nor a mortal price in souls of birds.

Her 'snow-mantled shadow' sings:

"Alas, my heart! No grief so great
As thinking on a happy state
In misery. Ah, dear is power
To female hearts! Oh, blissful hour

When Blanche and Flavia, joined with me,
Tri-feminine Directory,
Dispensed in latitudes below
The laws of flounce and furbelow;
And held on bird and beast debate,
What lives should die to serve our state!
We changed our statutes with the moon,
And oft in January or June,
At deep midnight, we would prescribe
Some furry kind, or feathered tribe.
At morn, we sent the mandate forth;
Then rose the hunters of the North:
And all the trappers of the West
Bowed at our feminine behest.
Died every seal that dared to rise
To his round air-hole in the ice;
Died each Siberian fox and hare
And ermine trapt in snow-built snare.
For us the English fowler set
The ambush of his whirling net;
And by green Rother's reedy side
The blue kingfisher flashed and died.
His life for us the seamew gave
High upon Orkney's lonely wave;
Nor was our queenly power unknown
In Iceland or by Amazon;
For where the brown duck stripped her breast
For her dear eggs and windy nest,
Three times her bitter spoil was won
For woman; and when all was done,
She called her snow-white piteous drake,
Who plucked his bosom for our sake."

126. "See 'Hartwig's Polar World' for the manner of taking Eiderdown."--Once more, we have thus much of author's note, but edition and page not specified, which, however, I am fortunately able to supply. Mr. Hartwig's miscellany being a favorite--what can I call it, sand-hill?--of my own, out of which every now and then, in a rasorial manner, I can scratch some savory or useful contents;--one or two, it may be remembered, I collected for the behoof of the Bishop of Manchester, on this very subject, (*Contemporary Review*, Feb. 1880); and some of Mr.

Hartwig's half-sandy, half-soppy, political opinions, are offered to the consideration of the British workman in the last extant number of 'Fors.' Touching eider ducks, I find in his fifth chapter--on Iceland--he quotes the following account, by Mr. Shepherd, of the shore of the island of 'Isafjardarjup'--a word which seems to contain in itself an introduction to Icelandic literature:--

127. "The ducks and their nests were everywhere, in a manner that was quite alarming. Great brown ducks sat upon their nests in masses, and at every step started up from under our feet. It was with difficulty that we avoided treading on some of the nests. The island being but three-quarters of a mile in width, the opposite shore was soon reached. On the coast was a wall built of large stones, just above the high-water level, about three feet in height, and of considerable thickness. At the bottom, on both sides of it, alternate stones had been left out, so as to form a series of square compartments for the ducks to make their nests in. Almost every compartment was occupied; and, as we walked along the shore, a long line of ducks flew out one after another. The surface of the water also was perfectly white with drakes, who welcomed their brown wives with loud and clamorous cooing. When we arrived at the farmhouse, we were cordially welcomed by its mistress. The house itself was a great marvel. The earthen wall that surrounded it and the window embrasures were occupied by ducks. On the ground, the house was fringed with ducks. On the turf-slopes of the roof we could see ducks; and a duck sat in the scraper.

"A grassy bank close by had been cut into square patches like a chess-board, (a square of turf of about eighteen inches being removed, and a hollow made,) and all were filled with ducks. A windmill was infested, and so were all the out-houses, mounds, rocks, and crevices. The ducks were everywhere. Many of them were so tame that we could stroke them on their nests; and the good lady told us that there was scarcely a duck on the island which would not allow her to take its eggs without flight or fear."

128. But upon the back of the canvas, as it were, of this pleasant picture--on the back of the leaf, in his book, p. 65,--this description being given in p. 66,--Doctor Hartwig tells us, in his own peculiar soppy and sandy way--half tearful, half Dryasdusty, (or may not we say--it sounds more Icelandic--'Dry-as-sawdusty,') these less cheerful facts. "The eiderdown is easily collected, as the birds are quite tame. The female having laid five or six pale greenish-olive eggs, in a nest thickly lined with her beautiful down, the collectors, after carefully removing the bird,

rob the nest of its contents; after which they replace her. She then begins to lay afresh--though this time only three or four eggs,--and again has recourse to the down on her body. But her greedy persecutors once more rifle her nest, and oblige her to line it for the third time. Now, however, her own stock of down is exhausted, and with a plaintive voice she calls her mate to her assistance, who willingly plucks the soft feathers from his breast to supply the deficiency. If the cruel robbery be again repeated, which in former times was frequently the case, the poor eider-duck abandons the spot, never to return, and seeks for a new home where she may indulge her maternal instinct undisturbed by the avarice of man."

129. Now, as I have above told you, these two statements are given on the two sides of the same leaf; and the reader must make what he may of them. Setting the best of my own poor wits at them, it seems to me that the merciless abstraction of down is indeed the usual custom of the inhabitants and visitors; but that the 'good lady,' referred to by Mr. Shepherd, manages things differently; and in consequence we are presently farther told of her, (bottom of p. 65,) that "when she first became possessor of the island, the produce of down from the ducks was not more than fifteen pounds weight in the year; but under her careful nurture of twenty years it had risen to nearly one hundred pounds annually. It requires about one pound and a half to make a coverlet for a single bed, and the down is worth from twelve to fifteen shillings per pound. Most of the eggs are taken and pickled for winter consumption, one or two only being left to hatch."

But here, again, pulverulent Dr. Hartwig leaves us untold who 'consumes' all these pickled eggs of the cooing and downy-breasted creatures; (you observe, in passing, that an eider-duck coos instead of quacking, and must be a sort of Sea-Dove,) or what addition their price makes to the good old lady's feather-nesting income of, as I calculate it, sixty to seventy-five pounds a year,--all her twenty years of skill and humanity and moderate plucking having got no farther than that. And not feeling myself able, on these imperfect data, to offer any recommendations to the Icelandic government touching the duck trade, I must end my present chapter with a rough generalization of results. For a beginning of which, the time having too clearly and sadly come for me, as I have said in my preface, to knit up, as far as I may, the loose threads and straws of my raveled life's work, I reprint in this place the second paragraph of the chapter on Vital Beauty in the second volume of 'Modern Painters,' premising, however, some few necessary words.

130. I intended never to have reprinted the second volume of 'Modern Painters'; first, because it is written in affected imitation of Hooker, and not in my own proper style; and, secondly, yet chiefly, because I did not think the analytic study of which it mainly consists, in the least likely to be intelligible to the general student, or, therefore, profitable to him. But I find now that the 'general student' has plunged himself into such abysses, not of analytic, but of dissolytic,--dialytic--or even diarrhoeic--lies, belonging to the sooty and sensual elements of his London and Paris life, that, however imperfectly or dimly done, the higher analysis of that early work of mine ought at least to be put within his reach; and the fact, somehow, enforced upon him, that there were people before *he* lived, who knew what 'aesthesis' meant, though they did not think that pigs' flavoring of pigs'-wash was ennobled by giving it that Greek name: and that there were also people before his time who knew what vital beauty meant, though they did not seek it either in the model-room, or the Parc aux Cerfs.

Therefore, I will republish (D.V.) the analytic parts of the second volume of 'Modern Painters' as they were written, but with perhaps an additional note or two, and the omission of the passages concerning Evangelical or other religious matters, in which I have found out my mistakes.

131. To be able to hunt for these mistakes, and crow over them, in the original volume, will always give that volume its orthodox value in sale catalogues, so that I shall swindle nobody who has already bought the book by bringing down its price upon them. Nor will the new edition be a cheap one--even if I ever get it out, which is by no means certain. Here, however, at once, is the paragraph above referred to, quite one of the most important in the book. The reader should know, preparatorily, that for what is now called 'aesthesis,' *I* always used, and still use, the English word 'sensation'--as, for instance, the sensation of cold or heat, and of their differences;--of the flavor of mutton and beef, and their differences;--of a peacock's and a lark's cry, and their differences;--of the redness in a blush, and in rouge, and their differences;--of the whiteness in snow, and in almond-paste, and their differences;--of the blackness and brightness of night and day, or of smoke and gaslight, and their differences, etc., *etc.* But for the Perception of Beauty, I always used Plato's word, which is the proper word in Greek, and the only possible *single* word that can be used in any other language by any man who understands the subject,--'Theoria,'--the Germans only having a term parallel to it, 'Anschauung,' assumed to be its equivalent in p. 22 of the

old edition of 'Modern Painters,' but which is not its real equivalent, for Anschauung does not (I believe) *include* bodily sensation, whereas Plato's Theoria does, so far as is necessary; and mine, somewhat more than Plato's. "The first perfection," (then I say, in this so long in coming paragraph) of the theoretic faculty, "is the kindness and unselfish fullness of heart, which receives the utmost amount of pleasure from the happiness of all things. Of which in high degree the heart of man is incapable; neither what intense enjoyment the angels may have in all that they see of things that move and live, and in the part they take in the shedding of God's kindness upon them, can we know or conceive: only in proportion as we draw near to God, and are made in measure like unto Him, can we increase this our possession of charity, of which the entire essence is in God only. But even the ordinary exercise of this faculty implies a condition of the whole moral being in some measure right and healthy, and to the entire exercise of it there is necessary the entire perfection of the Christian character; for he who loves not God, nor his brother, cannot love the grass beneath his feet, and the creatures which live not for his uses, filling those spaces in the universe which he needs not; while, on the other hand, none can love God, nor his human brother, without loving all things which his Father loves; nor without looking upon them, every one, as in that respect his brethren also, and perhaps worthier than he, if, in the under concords they have to fill, their part be touched more truly. It is good to read of that kindness and humbleness of S. Francis of Assisi, who never spoke to bird or cicala, nor even to wolf and beast of prey, but as his brother; and so we find are moved the minds of all good and mighty men, as in the lesson that we have from the mariner of Coleridge, and yet more truly and rightly taught in the Hartleap Well:--

'Never to blend our pleasure, or our pride,
With sorrow of the meanest thing that feels.'

And again in the White Doe of Rylstone, with the added teaching, that anguish of our own

'Is tempered and allayed by sympathies,
Aloft ascending, and descending deep,
Even to the inferior kinds;'

so that I know not of anything more destructive of the whole theoretic faculty, not to say of the Christian character and human intellect, than those accursed sports, in which man makes of himself, cat, tiger, serpent, chaetodon, and alligator in one; and gathers into one continuance of cruelty, for his amusement, all the devices that brutes sparingly, and at intervals, use against each other for their necessities."

132. So much I had perceived, and said, you observe, good reader, concerning S. Francis of Assisi, and his sermons, when I was only five-and-twenty,--little thinking at that day how, Evangelical-bred as I was, I should ever come to write a lecture for the first School of Art in Oxford in the Sacristan's cell at Assisi,[25] or ever--among such poor treasures as I have of friends' reliquaries--I should fondly keep a little 'pinch' of his cloak.

Rough cloak of hair, it is, still at Assisi; concerning which, and the general use of camels' hair, or sackcloth, or briars and thorns, in the Middle Ages, together with seal-skins (not badgers'), and rams' skins dyed gules, by the Jews, and the Crusaders, as compared with the use of the two furs, Ermine and Vair, and their final result in the operations of the Hudson's Bay Company, much casual notice will be found in my former work. And now, this is the sum of it all, so far as I can shortly write it.

There is no possibility of explaining the system of life in this world, on any principle of *conqueringly* Divine benevolence. That piece of bold impiety, if it be so, I have always asserted in my well-considered books,--I considering it, on the contrary, the only really pious thing to say, namely, that the world is under a curse, which we may, if we will, gradually remove, by doing as we are bid, and believing what we are told; and when we are told, for instance, in the best book we have about our own old history, that "unto Adam also, and to his wife, did the Lord God make coats of skins, and clothed them," we are to accept it as the best thing to be done under the circumstances, and to wear, if we can get them, wolf skin, or cow skin, or beaver's, or ermine's; but not therefore to confuse God with the Hudson's Bay Company, nor to hunt foxes for their brushes instead of their skins, or think the poor little black tails of a Siberian weasel on a judge's shoulders may constitute him therefore a Minos in matters of retributive justice, or an AEacus in distributive, who can at once determine how many millions a Railroad Company are to make the public pay for not granting them their exclusive business by telegraph.

133. And every hour of my life, since that paragraph of 'Modern Painters' was written, has increased, I disdain to say my *feeling*, but say, with fearless decision, my *knowledge*, of the bitterness of the curse, which the habits of hunting and 'la chasse' have brought upon the so-called upper classes of England and France; until, from knights and gentlemen,

they have sunk into jockeys, speculators, usurers, butchers by battue; and, the English especially, now, as a political body, into what I have called them in the opening chapter of 'The Bible of Amiens,'--"the scurviest louts that ever fouled God's earth with their carcasses."

The language appears to be violent. It is simply brief, and accurate. But I never meant it to remain without justification, and I will give the justification here at once.

Take your Johnson, and look out the adjective Scurvy, in its higher or figurative sense.

You find the first quotation he gives is from 'Measure for Measure,' spoken of the Duke, in monk's disguise:

"I know him for a man divine and holy;

Not scurvy, nor a temporary meddler."

In which passage, Shakspeare, who never uses words in vain, nor with a grain less than their full weight, opposes the divineness of men, or their walking with God, to the scurviness of men, or their wallowing with swine; and again, he opposes the holiness of men,--in the sense of "Holy--harmless, undefiled," and more than that, helpful or healthful in action--to the harmful and filthy action of temporary meddlers, such as the hanging of seventeen priests before breakfast, and our profitable military successes, in such a prolonged piece of 'temporary meddling' as the Crimean war.

134. But, secondly, if you look down Johnson's column, you will find his last quotation is not in the higher or figurative, but the lower and literal sense, from Swift, to the effect that "it would be convenient to prevent the excess of drink, with that scurvy custom of taking tobacco." And you will also find, if you ever have the sense or courage to look the facts of modern history in the face, that those two itches, for the pot and the pipe, have been the roots of every other demoralization of the filthiest and literally 'scurviest' sort among *all* classes;--the dirty pack of cards; the church pavement *running* with human saliva,--(I have seen the spittings in ponds half an inch deep, in the choir of Rouen cathedral); and the entirely infernal atmosphere of the common cafes and gambling-houses of European festivity, infecting every condition of what they call 'aesthesis,' left in the bodies of men, until they cannot be happy with the pines and pansies of the Alps, until they have mixed tobacco smoke with the scent of them; and the whole concluding in the endurance--or even enjoyment--of the most squalid conditions of filth in our capital cities, that have ever been yet recorded, among the disgraces of mankind.

135. But, thirdly, Johnson's central quotation is again from 'Measure for Measure':--

"He spoke *scurvy* and *provoking* terms against your honor."

The debates in the English House of Commons, for the last half-century, having consisted virtually of nothing else!

I next take the word 'lout,' of which Johnson gives two derivations for our choice: it is either the past participle of 'to lower, or make low;' a lowed person, (as our House of Lords under the direction of railway companies and public-house keepers); or else--and more strictly I believe in etymology--a form of the German 'leute,' 'common people.' In either case, its proper classical English sense is given by Johnson as "a mean, awkward fellow; a bumpkin, a clown."

Now I surely cannot refer to any general representation of British society more acceptable to, and acknowledged by, that society, than the finished and admirably composed drawings of Du Maurier in *Punch* which have become every week more and more consistent, keen, and comprehensive, during the issues of the last two years.

I take three of them, as quite trustworthy pictures, and the best our present arts of delineation could produce, of the three Etats, or representative orders, of the British nation of our day.

Of the Working class, take the type given in Lady Clara Robinson's garden tea party, p. 174, vol. 79.

Of the Mercantile class, Mr. Smith, in his drawing-room after dinner, p. 222, vol. 80.

And of the Noblesse, the first five gentlemen on the right (spectator's right) of the line, in the ball at Stilton House, (July 3d, 1880).

136. Of the manner or state of lout, to which our manufacturing prosperity has reduced its artisan, as represented in the first of these frescoes, I do not think it needful to speak here; neither of the level of sublime temperament and unselfish heroism to which the dangers of commercial enterprise have exalted Mr. Smith. But the five consecutive heads in the third fresco are a very notable piece of English history, representing the polished and more or less lustrous type of lout; which is indeed a kind of rolled shingle of former English noblesse capable of nothing now in the way of resistance to Atlantic liberalism, except of getting itself swept up into ugly harbor bars, and troublesome shoals in the tideway.

And observe also, that of the three types of lout, whose combined chorus and tripudiation leads the present British Constitution its devil's dance, this last and smoothest type is also the dullest. Your operative lout cannot indeed hold his cup of coffee with a grace, or possess himself of a biscuit from Lady Clara's salver without embarrassment; but, in his own mill, he can at least make a needle without an eye, or a nail without a head, or a knife that won't cut, or something of that sort, with dexterity. Also, the middle class, or Smithian lout, at least manages his stockbroking or marketing with decision and cunning; knows something by eye or touch of his wares, and something of the characters of the men he has to deal with. But the Ducal or Marquisian lout has no knowledge of anything under the sun, except what sort of horse's quarters will carry his own, farther weighted with that smooth block or pebble of a pow; and no faculty under the sun of doing anything, except cutting down the trees his fathers planted for him, and selling the lands his fathers won.

137. That is indeed the final result of hunting and horse-racing on the British landlord. Of its result on the British soldier, perhaps the figures of Lord George Sackville at the battle of Minden, and of Lord Raglan at the battle of Alma, (who in the first part of the battle did not know where he was, and in the second plumed himself on being where he had no business to be,) are as illustrative as any I could name; but the darkest of all, to my own thinking, are the various personages, civil and military, who have conducted the Caffre war to its last successes, of blowing women and children to death with dynamite, and harrying the lands of entirely innocent peasantry, because they would not betray their defeated king.

138. Of the due and noble relations between man and his companion creatures, the horse, dog, and falcon, enough has been said in my former writings--unintelligible enough to a chivalry which passes six months of its annual life in Rotten Row, and spends the rents of its Cumberland Hills in building furnaces round Furness Abbey; but which careful students either of past knighthood, or of future Christianity, will find securely and always true. For the relations between man and his beast of burden, whether the burden be himself or his goods, become beautiful and honorable, just in the degree that both creatures are useful to the rest of mankind, whether in war or peace. The Greeks gave the highest symbol of them in the bridling of Pegasus for Bellerophon by Athena; and from that myth you may go down to modern times--understanding, according to your own sense and dignity, what all prophecy, poetry, history, have told you--of the horse whose neck is clothed with thunder, or the ox who treadeth out the corn--of Joseph's chariot, or of Elijah's--of Achilles and Xanthus--Herminius and Black Auster--down to Scott and Brown Adam--or Dandie Dinmont and Dumple. That pastoral one is, of

all, the most enduring. I hear the proudest tribe of Arabia Felix is now reduced by poverty and civilization to sell its last well-bred horse; and that we send out our cavalry regiments to repetitions of the charge at Balaclava, without horses at all; those that they can pick up wherever they land being good enough for such military operations. But the cart-horse will remain, when the charger and hunter are no more; and with a wiser master.

"I'll buy him, for the dogs shall never
Set tooth upon a friend so true;
He'll not live long; but I forever
Shall know I gave the beast his due.
Ready as bird to meet the morn
Were all his efforts at the plow;
Then the mill-brook--with hay or corn,
Good creature! how he'd spatter through.
I left him in the shafts behind,
His fellows all unhook'd and gone;
He neigh'd, and deemed the thing unkind;
Then, starting, drew the load alone.

Half choked with joy, with love, and pride,
He now with dainty clover fed him;
Now took a short, triumphant ride,
And then again got down, and led him."

139. Where Paris has had to lead *her* horses, we know; and where London had better lead hers, than let her people die of starvation. But I have not lost my hope that there are yet in England Bewicks and Bloomfields, who may teach their children--and earn for their cattle--better ways of fronting, and of waiting for, Death.

Nor are the uses of the inferior creatures to us less consistent with their happiness. To all that live, Death must come. The manner of it, and the time, are for the human Master of them, and of the earth, to determine--not to his pleasure, but to his duty and his need.

In sacrifice, or for his food, or for his clothing, it is lawful for him to slay animals; but not to delight in slaying any that are helpless. If he choose, for discipline and trial of courage, to leave the boar in Calydon, the wolf in Taurus, the tiger in Bengal, or the wild bull in Aragon, there is forest and mountain wide enough for them: but the inhabited world in sea and land should be one vast unwalled park and treasure lake, in which its flocks of sheep, or deer, or fowl, or fish, should be tended and dealt with, as best may multiply the life of all Love's Meinie, in strength, and use, and peace.

THE END

Made in United States
Cleveland, OH
06 November 2026

47647920R10051